ALLIED MINESWEEPING IN WORLD WAR 2

ALLIED MINESWEEPING IN WORLD WAR 2

Peter Elliott

NAVAL INSTITUTE PRESS

Library of Congress Catalog Card No 78-61581
ISBN 0-87021-904-9.

Printed in Great Britain.

Published and distributed in the United States of America
by the Naval Institute Press, Annapolis, Maryland 21402.

Contents

Preface

The object of this book is to provide a specialised but interesting history of British minesweeping in World War 2, and equally to cover the great American and Canadian minesweeping achievements of that war.

In addition to the general narrative, there are detailed accounts of the Normandy landings, of the clearance of the River Scheldt, and of the US Navy's Pacific island-hopping assaults, with special accounts of the operations around Borneo and Okinawa. The story of the pressure mines, and of the special sweeps devised against them, is told in detail for the first time, and there is a description of 'assault sweeping', a technique devised for the many landing operations.

The early experimental sweeps produced by the British against the new German ground mines are well covered, and there are surprises too—the moored contact mine, for example, developed in World War 1 remained the single biggest mining hazard right through to 1945, for both the British and the Americans.

But how the ground mines changed the sweeping methods! Here is a quote from a British guide for sweeping mixed fields in 1945—and this was only one part of the operation:

'Sweep with Modified L and third electrode in "R" or "S" formation for at least ten passes over the area. Explosive sweep should be used on first few passes, and SA oscillator towed in the first sub-division to enter the area; one SA sweep should be used for each 400 yards of swept path per lap.'

A long way from the original single wire sweep of the trawlers!

To help the reader follow a narrative which includes some technical terms, the sweeps, sweeping formations, and new minesweeper designs are described chronologically, where they first appear. Also, there is a glossary at the end of the book. Most of the information is based on the British experience, but my American friends have reminded me that the biggest impetus in inventing new types of sweep lay in the mining campaign in northern Europe.

The book is based on the author's extensive research, and the bibliography will point the interested reader in the right direction. In England, a very large number of thick files containing the original British plans and action reports are now available in the Public Records Office (including, for example, the story of the pressure mines), and these include copies of many original United States Navy reports for the Pacific. The National Archives in Washington has a magnificent collection of photographs of American sweepers. The cooperation of the naval and archive authorities in both countries, as in the other countries named in the acknowledgements, has been greatly valued.

Even in a book of this size, space has been a big problem. Many stories have had to be left out, as have analyses of sweepers in various war theatres, and class lists of the sweepers themselves. Similarly, minelaying and Axis minesweeping could not have been included in this one volume, without making it a jumble of facts and figures.

However, I hope that I have given the reader, be he historian or old naval hand, a good picture of a side of World War 2 which has been publicised little enough but was both important and colourful.

Peter Elliott

The Ditty of the Pressure Minesweepers

Humour was an important element in the sweepers' dangerous work. At the height of the 'oyster' mining campaign in Europe, this poem appeared in the weekly official British mine sweeping summary.

I needn't tell you all the rest—
The Germans did their level best
To find an antidote, and failed.
But then we were ourselves assailed
With Aryan oysters. Things looked blue
Till Surgeon Captain Spleen came through
With something that he'd learned when he
Was in Peru in '83.
This was that oysters worship stout;
They will not close when it's about,
And we could beat the German mine
By mixing Porter with the brine.
Native *was fitted out once more,*
And duly cruised about offshore
Sluicing good liquor on the sea.
This acted most efficiently,
And even had the Hun found out
The secret, as he can't make stout,
He couldn't have made use of it.
His ships continued to be hit,
And ours were perfectly all right.
That's how we won the war. Good night!

Acknowledgements

Grateful thanks to the following, for providing access to research facilities, information, or photographs:

United Kingdom
The Controller, HM Stationery Office (for permission to reproduce the poem opposite); David Lyon; Duncan Haws (author of the fine new series of books *Merchant Fleets in Profile*) for the silhouette drawings; the Imperial War Museum; the following departments of the Ministry of Defence—Admiralty Underwater Weapons Establishment, Portland; the Director General Ships, Bath; the Naval Historical Branch, London; the Naval Home Division, London and Bath; the Royal Naval Armament Depot, Priddy's Hard, Portsmouth; HMS *Vernon*, Portsmouth. Also the National Maritime Museum, Greenwich; Antony Preston, London; the Public Records Office, London; P. A. Vicary, Cromer, Norfolk; Wright & Logan, Portsmouth.

United States of America
The Department of the Navy—Naval Historical Center, Washington; the Office of Information, Washington; Lt Cdr Arnold S. Lott, USN (Ret); the National Archives—Audio-Visual Archives Division; *Warship International*.

Australia
The RAN Historical Section, the Department of Defence, Melbourne; Lt Cdr R. P. Hall, VRD, RANR, Melbourne.

Canada
The Director of History, the Department of National Defence, Ottawa; the National Photography Collection, Public Archives of Canada, Ottawa.

South Africa
Vice-Admiral J. C. Walters, SM, Chief of the Navy.

Glossary

Acoustic mine A mine actuated by sound from the propellers of a passing ship.
Actuations The number of times a mine is actuated by the necessary influence (magnetic, acoustic, pressure) before the arming delay is satisfied, and fires the mine.
AM American fleet minesweeper class.
AMc American coastal minesweeper class.
Ampere A measure of electrical influence, or current, used to measure, for instance, the power fed into an LL sweep.
APD American destroyer escort, converted to a fast troop transport, and carrying four LCVPs.
Arming delay A mechnanical delay mechanism, built into ground influence mines, by which the mine is not armed until the set number of hours or days have passed.
Asdic Submarine detection equipment in anti-submarine vessels, also fitted in fleet sweepers, and latterly used to locate mines.
'A' sweep Wire sweep, streamed between the sterns of two or more sweepers, and used to sweep moored mines. Seldom used in World War 2, as sweepers' freedom of manoeuvre was severely restricted under air attack.

BAM 'British American Minesweeper', or British AM; a unit of the American 'Raven' Class transferred to the Royal Navy under Lend-Lease.
BYMS 'British Yard Minesweeper', a unit of the American YMS class transferred to the Royal Navy under Lend-Lease.
Boyes rifle Army anti-tank rifle, mounted near the bridge of a British sweeper, and used as a high-velocity rifle for sinking mines.

Catenary sweep A wire or electric sweep, streamed in a loop between the sterns of two or more sweepers.
Clearance sweep A sweep by a unit or group of sweepers, designed to clear a defined area, or an enemy minefield already located by an exploratory or searching sweep.
Coarse A mine setting by which the magnetic, acoustic, or pressure signature required to fire the mine needs to be a strong one. The opposite was a fine setting.
Coastal sweeper A minesweeper designed for operations in coastal waters or harbours. They frequently voyaged all over the world.
Combination mine A mine incorporating two or more different influences

(magnetic, acoustic or pressure), each of which needs to be actuated by a passing ship or sweep before the mine will fire.

Communicating channels Narrow channels swept through a large mined area, and marked by dans, for the safe movement of sweepers.

Concrete barge A displacement sweep, specially constructed in the UK for use against pressure mines.

CTL Constructive Total Loss, when a sweeper is severely damaged, and though brought back to port by her crew, is not worth repairing. This type of damage became frequent when influence mines were used in coastal waters.

Cutter A cutting device fitted to a wire sweep close to the outer end, to sever the moorings of mines swept by the sweep.

Dan-buoy Light spar buoy, adapted from fishing practice, widely used for marking the channel swept by minesweepers.

Danlayer A ship laying or lifting dan-buoys. In the Royal Navy, these were initially requisitioned trawlers, later they were specially converted new construction trawlers, and on occasion, fleet sweepers or YMSs and MMSs.

De-gaussing The neutralisation of the magnetic field of a ship, by passing an electric current through a cable running round the outside of the hull.

DMS American destroyer minesweeper. A large class, widely used in assault operations.

Displacement sweep A variety of sweep types, used against pressure mines. The types included Stirling craft, Egg Crates, Concrete barges, 'guinea-pigs' and other experimental types.

Displacer A device with a cylindrical body, streamlined at its outer ends, and fitted with a depressing plane underneath, for use as a small kite with light Oropesa sweeps.

Drifter A small fishing vessel, usually built of wood, which, when fishing, streamed its nets and drifted downwind from them, as opposed to trawling.

E-boat German or Italian motor torpedo boat. 'E-boat' was a generic term used by the Allies.

Egg Crate A steel-built displacement sweep, produced quickly by the Americans and British when German pressure mines appeared off Normandy in June 1944.

Exploratory sweep A routine or preliminary sweep, usually of a war channel, to search for any mines laid since the last sweep.

Fine A fine setting on an influence mine (as opposed to a coarse setting), needing only a slight reaction of the mine to fire it.

Fire In minesweeping terms, to explode a mine.

Fast sweeper The pre-1939 term for a fleet sweeper in the Royal Navy.

Fleet sweeper The largest type of minesweeper (AM in the United States Navy), later called an Ocean Minesweeper (MSO).

Float A steel or wooden float, used to mark the end of a wire or electric cable sweep.

Floating mine A moored mine which has broken away from its moorings, or been cut by a sweeper's wire.

Flotilla Royal Navy term for a group of ships. In the United States Navy this was termed a sweep unit.

Four-piper A destroyer minesweeper of the United States Navy, of a large class built in World War 1.

Four-stacker Another term for a four-piper.

Frequency The frequency of a sound source, as in an acoustic mine, measured in cycles per second.
Gauss A unit of electro-magnetic density, used to measure, eg, the magnetism of a steel ship.
Ground mine A mine laid on the seabed, and so of the influence type—magnetic, acoustic or pressure.
Group A group of minesweepers—usually applied to a group of trawlers in the Royal Navy.
Guinea-pig A merchantman, destroyer or large landing craft (usually in a damaged state) used to prove a swept channel free of pressure mines, by passing through it just to see.
HDML Harbour Defence Motor Launch, a very large class of wood-built harbour craft in the Royal Navy. A number were fitted as light Oropesa sweepers.
HSMS High speed destroyer minesweep—a Royal Navy minesweep fitted pre-1939 on the sterns of a number of fleet destroyers, and notable by two prominent davits. Not often used in wartime, although successful in the Mediterranean.
Hydrographic unit A group of trawlers, frigates or survey vessels used to chart an area recently captured from the enemy. Used by the Royal Navy in the River Scheldt, and by the USN and RAN extensively in the Pacific.
Influence mine A ground mine, using acoustic or pressure influences to fire it.
Inshore sweeper A small sweeper, usually built of wood, used for minesweeping in shallow water. Not larger than an MMS or YMS.
Kamikaze Japanese aircraft flown by a suicide pilot and crash-landed on warships in the Pacific, causing widespread damage.
Kango hammer A British pneumatic road drill, used in the main acoustic sweep in the Royal Navy.
Katie Mine A contact mine used by the Germans in 1944 and 1945, mounted on a concrete base with an iron top, and fired by contact with snaglines floating on the surface.
Kite Together with the Otter, the Kite kept the wire of the Oropesa sweep at the required sweeping depth below the surface of the sea.
LAA An LL magnetic sweep, adapted for use by sweepers in fresh water or rivers. In Europe, usually operated by a BYMS, MMS, MFV or ML.
Lap A flotilla of sweepers, sweeping in formation along a line of dan buoys, complete a lap, or swathe, through the minefield.
LCA Landing Craft Assault, a small craft which could operate a light rope Oropesa sweep.
LCT Landing Craft Tank, a steel craft used to carry tanks to landing beaches, and which could carry a light rope Oropesa sweep on its way in.
LCVP Landing Craft Vehicles and Personnel—much like an LCA.
Lift A term used in sweeping to denote the exploding of a ground mine.
LL The main magnetic sweep, consisting of a pair of insulated and buoyant electric cables towed astern of the sweeper, activated by a generator or batteries on the sweeper.
LST Landing Ship Tank, a much larger ship than the LCT, although used in the same operation. LSTs did not tow sweeps.
Magnetic mine A ground mine operated by the magnetism in a steel ship's hull.
Maintenance sweeping The maintaining by routine sweeping of swept channels

round a coast, or in the assault area near landing beaches.

MFV Motor Fishing Vessel, a wooden fishing-type sweeper operated by the Royal Navy, especially in rivers and harbours. Its deep draught put it at risk in many situations.

Mine location A technique first developed towards the end of World War 2 and subsequently, to locate mines on the seabed, and then to neutralise them.

ML Motor Launch, a large class of wooden launch in the Royal and Commonwealth Navies, used inshore. They made excellent light Oropesa sweepers, and were much used at Malta in 1942–43, and in assault operations.

MMS Motor Minesweeper, a large class of wooden coastal minesweeper in the Royal Navy, and the equivalent of the American YMS.

Moored contact mine The mine with which all navies started World War 2, and which continued in full use through to 1945.

M/SML Minesweeping Motor Launch, the sweeping version of the ML.

MTB Motor Torpedo Boat, in the Royal Navy.

NF Channel The swept channel from the North Foreland, in south-east England, to the entrance of the River Scheldt, leading to Antwerp, a channel much disputed in 1944–45.

OA Overall, as in the length of a ship. Also o/a.

Ocean sweeper Post-1945 term for a fleet sweeper, or AM.

Oropesa sweep The name in all Allied navies for the standard wire sweep, named after the RN trawler of that name in which the sweep was first developed in 1919.

Otter With the Kite, the Otter kept the wire sweep down below the surface of the sea, to the required sweeping depth.

Overlap mine A mine developed in 1944–45 to frustrate the sweepers. It is described in the narrative at that point.

Oyster The British code-name for the German pressure mine, first used operationally in June 1944 at the Normandy beach-head.

Paddle sweeper A steamer propelled by paddles, with a shallow draught, and used by the Royal Navy for minesweeping in the early years of World War 2.

PDM Pulse Delay Mechanism—a device which required a pre-set number of actuations (qv) before the mine which contained it would fire.

Pressure mine A ground mine operated by the pressure waves of a ship passing overhead.

Quarter line The usual formation for a unit of minesweepers using the wire Oropesa sweep. Each ship steamed in the safe water inside the Oropesa float of the next ship ahead.

RAN Royal Australian Navy.

RCN Royal Canadian Navy.

RN Royal Navy.

RNZN Royal New Zealand Navy.

Running free A sweeper steaming without her sweeps streamed, and so moving freely and faster through the water.

SA An acoustic sweep.

SANF South African Naval Forces.

SC Submarine Chaser, a USN wooden ship type.

Searching sweep Routine sweep, of, eg, a war channel.

Sensitivity The fine or coarse setting of a mine, in terms of the degree of influence required to fire it.

SGB Steam Gun Boat, an RN Coastal Forces type used for magnetic and pressure sweeping in 1944–45.

Ship counter A ratchet type device in a ground mine, by which the required number of actuations (by ships passing overhead) is counted before the mine will fire.

Sinker The concrete or iron weight, on the lower end of the mooring wire of a moored mine or a dan-buoy, by which it is anchored to the seabed.

SKD Station Keeping Distance, the distance between ships observed by a group or unit of sweepers.

Skid A wooden dumb lighter, with a magnetic coil mounted on top, used during the early days of magnetic sweeps in the Royal and United States Navies.

Sloop A Royal Navy ship type, which before 1939 carried minesweeping gear.

Snag Line A rope, floating on the surface of the sea, and connected at one end to the firing contact of a mine. Used extensively by the Germans in Europe in 1944–45.

Sound signature The sound waves caused by the propellers of a passing ship, and registered by an acoustic mine's firing unit.

Squadron A unit of minesweepers, usually of the AM type, in the United States Navy.

Statistical sweeping The observation from ships or from the shore, of mines laid by aircraft, usually at night in coastal or river areas.

Sterilise A mine could sterilise itself by being pre-set to become safe, or to sink to the seabed, at a pre-arranged date or time.

'Stirling' craft Two steel 'Stirling' craft were built by the Royal Navy in 1943–44, as sweepers of pressure mines.

Stream To stream a sweep is to pass it out over the stern of a sweeper (SA sweeps could be lowered over the bow or from amidships).

Sweep deck The deck on the stern of the sweeper where most of the minesweeping gear is located (also quarterdeck in the RN, fantail in the USN).

Sweep obstructor A device attached to a moored mine to cut or obstruct the sweep wire of a minesweeper.

Tail An LL tail, the magnetic sweep streamed astern of the sweeper.

Task force or unit American Navy terms for groups of minesweepers.

Trawler A fishing vessel, trawling for fish with a large net towed astern of it.

USN United States Navy.

War channel The swept channel maintained along a coast, or in the approach to a harbour.

Whaler A small steel vessel, used for hunting whales. Used as sweepers during the war. Distinguished by their very low foredecks, used for handling whales in peacetime.

Wiping A system for reducing the magnetism of a steel ship's hull, by raising and lowering electric cables charged with current around the outside of the hull.

Wire sweep The Oropesa sweep against moored mines.

YMS Yard Motor Minesweeper, the largest American war-built class of wooden coastal minesweeper.

'ZZ' craft Wooden lighters, originally built in the Middle East for magnetic sweeping up rivers, but later tested out as possible pressure sweepers.

Chapter 1

Sweeping between the wars, 1919–1939

First, one should examine the progress made in minesweeping techniques and in the construction of new types of sweeper between the two World Wars.

The Royal Navy needed to react to the approaching conflict earlier than the Americans did and, by the time the latter entered the war, the British already had two years' experience. The knowledge gained was readily passed on to the US Navy so that they could progress more quickly, once the green light was given to expand their minesweeping service.

Minesweeping in the Royal Navy

In Britain, the proximity of Germany and the gathering clouds of war in the 1930s ensured that, in spite of cuts in warship building budgets, minesweeping skills and technology progressed quietly but steadily—though almost entirely in the field of wire sweeping. For it was the Admiralty's considered opinion, until early in 1939, that the chief danger to shipping in the shallow waters around the British Isles lay in the moored contact mine, and it was to meet this weapon that the Royal Navy's minesweeps were specifically designed. First we must consider what happened to the wire sweep, not as a curio surviving from World War 1, but as a device which, right through World War 2, continued to play a major part in mine warfare in every theatre of operations.

Indeed, looking back in 1946, the Admiralty in London reflected with awe that the plain, moored contact mine had still been effective in 1945, if only because the anti-sweeping devices produced by the Germans—obstructors, and ever-heavier chain moorings which the wire sweeps could not cut—were still requiring a major effort, even from the very large forces of new sweepers.

A further indication of this is that the Royal Navy's war-built fleet sweepers were very largely employed on wire sweeping right through to 1944, and it was not until then, when the LL magnetic sweeps in these classes were used operationally in full force for the first time, that defects in the diesel generators fitted in some of the classes came to light.

A more comforting thought to the Admiralty in retrospect was that its wire sweep seemed to be up to 20 years ahead of that in any other navy in 1939, and even in 1945 was still considered to be the best. But before early 1939, the developmental thinking seemed to stop there, and very little work seems to have been done on the influence ground mine.

Improvements in the sweeps

The lessons of the 1919 clearance operations were certainly taken to heart in the

AMc 91 Accentor, *a unit of the coastal class in June 1945. This class was quickly built in the early war years for inshore work.*

next ten years. The original single-ship wire sweep, named the Oropesa after the trawler in which it was first tested, was clearly more seaworthy than the 'A' sweep, a wire towed between the sterns of two ships. The heavy-duty Mark I was first introduced for the minesweeping sloops produced at the end of World War 1; but it was found that only the 'Flower' Class sloops, and the inter-war fleet sweepers built ten years later, had the engine power to tow this device at a satisfactory operating speed, or the deck space needed aft to handle the bulky gear quickly.

A lighter sweep, the Mark II, was introduced for trawlers, but it did not work very well, since the otter board it needed also had a high water resistance. Thus in 1937 the Mark II* came out and, with the often successful compromise, combined the otter and float of the earlier equipment with the lighter sweep wire and kite wire of the later one. This worked out pretty well and gave a high sweeping speed to the paddle sweepers, the faster fishing trawlers and the new Admiralty-designed trawlers then beginning to appear.

A sensible-sized outfit of sweeping gear was also laid down for each class of sweeper, which avoided duplication. This allowed each ship to operate a double Oropesa sweep at the same time—a wire sweep out on each side.

Looking further ahead to the needs of war, another type of sweep was seen to be essential for the smaller drifters. This was later found to be ideal for the motor minesweepers, and American-built yard minesweepers.

Building up the British sweepers

We start with the big fleet of sweepers built up in 1919 for mine clearance. Some carried on during World War 2, and it is helpful to look at that great force, in understanding the sweepers with which the Royal Navy entered the war. These were the principal classes:

The old and the new. The paddle sweeper Sandown *(built for the Southern Railway in 1936) passes MMS 14 on her way out of harbour.*

'Flower' Class

Fourteen ships were completed as sweepers, for duty with the Grand Fleet. Many others were completed as anti-submarine escorts, for convoy duty—the earlier units of the 'Flower' Class corvettes completed in 1939–42 were often fitted for minesweeping work as well as escort duties, so carrying this honoured name forward. Very few of this famous class survived to 1939, and none was fit for use as a sweeper, so at this point they fade from our story.

'Hunt' Class

This was the last class of sloops built in the 1918 period, and as many as 21 of the originals were still in service in 1939. Although old, they were quite large and performed great service throughout World War 2; they were employed largely in the Mediterranean, having previously served both at home and in the Far East. The last two surviving fleet sweepers at Malta during the great siege of 1942 were of this class, and one flotilla survived to operate together in the forefront of the Normandy invasion in 1944—a notable achievement.

Their displacement was 730 tons, with an overall length of 220 feet, and a draught of ten feet. Originally all coal-fired, they remained so in a number of units through to 1945 and they were universally known as 'Smokey Joes'. With a fuel capacity of 186 tons, they had a range of 2,000 miles. Their sweeping capacity was similar to the earlier 'Flower' Class; one pair could sweep $2\frac{1}{4}$ square miles of water in an hour.

Paddle minesweepers

In 1916, an excellent Admiralty fleet of paddle sweepers had been built; the trawlers could only sweep at low speeds and could only go inshore safely at certain states of the tide, due to their fishing-style deep draught. The paddlers, however, had not only a good turn of speed, but also a very shallow draught and they gave excellent service.

For years before 1914, paddle steamers had been in service around the coasts of Britain in the summer months, carrying crowds of trippers on memorable adventures and, after 1918, over 70 of these paddlers carried on in this trade. When war came in 1939, these ships were swiftly requisitioned into naval service once again and, within weeks, five flotillas of them were at sea, taking the brunt of the wire clearance sweeping around the coasts. Forty ships made up these flotillas, and 23 of them dated from before World War 1.

In 1942, the need for efficient anti-aircraft vessels which could operate close inshore was great, and the shallow draught and wide decks of the paddlers made them ideal for this task. By that time, new sweepers were coming into service in numbers, so that most of them became Auxiliary A/A ships; but one flotilla of paddle sweepers operated together until 1945. Before then, they had paid their price in losses. Nine of them were lost as sweepers; and they were very active during the Dunkirk evacuation—four of those nine being lost off the beaches there.

They could stay at sea for up to six days at a stretch, extra coal and water storage being easily fitted. Their normal wire sweep was still the old 'A' variety, the wire towed between two ships and, to handle this, a large steel gallows was fitted over the stern. Some units of this class retained the gallows throughout their minesweeping service, but others swapped the gallows for the later twin-geared minesweeping davits, for handling the Oropesa floats.

Their draught was a surprising $6\frac{1}{2}$ feet, compared with the 14 feet of a trawler, and their sweeping speed was 9 knots, against the 5–6 knots of the fishing vessels. Their free speed was 15 knots, and they carried a crew of 50. Their overall length was 227 feet and, with their big paddle boxes included, their breadth was as much as 57 feet.

Trawlers and drifters

Several large Admiralty-designed classes had been built during World War 1, and had proved themselves to be excellent wire sweepers. Quite a number were still in Admiralty service in 1939, and hundreds more were still in commercial use and were requisitioned on the outbreak of war. In the astonishing figures of requisitioned trawlers, drifters, and whalers, which we shall see in Royal Navy service in 1939–45, there was a relatively large number of ships which had

St Kilda, *an 'Isles' Class trawler, laying dans. One can be seen in the water astern of her, with flag flying; she has many more on deck.*

originated in Admiralty designs many years earlier. In the paddle sweepers and drifters, the Admiralty had built up a rich heritage of soundly-designed vessels from World War 1, which made a significant contribution to the minesweeping forces available during the critical early years of World War 2. When arguments come forward about unpreparedness in 1939, this important fact is all too often forgotten.

The new minesweepers, 1930–39

By 1930, Royal Navy thinking on the tactical use of minesweepers was picking up speed. The operating roles of the three main types of sweeper were seen like this:

Fleet minesweepers

Were used for searching and clearing swept channels to fleet operating bases and defining the limits of enemy minefields and clearing them.

Fast paddle sweepers

Carried out exploratory or searching sweeps, especially close inshore, and cleared enemy minefields.

Trawlers and drifters

Were engaged in searching and keeping clear the coastal swept channels and approaches to commercial harbours and clearing discovered minefields, if fleet or fast sweepers were not available.

These roles, predictably, changed quite a bit when war experience became available. The fleet sweepers were indeed used near the fleet bases, but were also used extensively, as their numbers grew, in the clearance of swept channels and, from 1943 onwards, in assault operations.

The paddle sweepers were to disappear early in the war from the sweeping forces but the requisitioned trawlers, drifters and whalers operated in large numbers very much within their defined role.

These initial roles may be summarised like this:

Class	Average sweeping speed, in knots	Maximum draught	Limit of suitable sweeping weather
'Flower' Class sloops	10½	12½ feet	Force 5
'Bridgewater' Class sloops	10	10½ feet	Force 5
'Town' Class fast sweepers	9–10	10 feet	Force 4
Paddle sweepers	8–9	7 feet	Force 3–4
Trawlers	5–6	14–15 feet	Force 5

'Bridgewater' Class

Like the US Navy, the Royal Navy coasted along for some ten years after World War 1 with the minesweeping sloops of the 'Flower' and 'Hunt' Classes, completed at the end of the war, as the backbone of the minesweeping force. But 1928 saw the launching of the first of several new but small classes of sloop. These were good-looking ships, with three or four being built each year. This was partly done to keep the smaller warship building yards in business, with a nucleus of specialised warship-building experience, and partly to develop naval thinking about this class of ship.

For some years attempts were made to combine within one sloop hull all the three main requirements—anti-submarine escort, anti-aircraft protection and minesweeping. There had not been much progress in any of these since World War 1, and financial economy contributed to this approach.

Looking at these sloops, we can see that the hull design, the bridge and funnel, carry over to the wartime 'Black Swan' and 'Algerine' Classes. Each branched out into a specialised design, but the similarities cannot be mistaken. The tall masts and rigging also persisted in both, to be cut down once the need for clear fields of fire for the anti-aircraft guns became apparent.

The armament varied with the specialisation. These early sloops started with one or two single 4.7-inch guns forward and aft, but with no real anti-aircraft guns. The twin 4-inch mounting did not come until the later 'Black Swan' Class was in service, and not at all to the minesweeping ships. The boats were a real peacetime outfit, including the 14-foot sailing dinghy in its own davits—and this was carried over to the wartime 'Algerine' Class, as were the partly-planked upper decks.

The minesweeping equipment clearly took precedence over the anti-submarine depth charges in these early ships and, as there was not room for both, this was probably the main factor in the decision to split the classes between the three operational requirements. The large steam-driven minesweeping winch took pride of place on the quarterdeck of these ships, the old gallows for the 'A' sweep of the 'Hunt' Class giving way to the efficient twin davits, one on each quarter, for the Oropesa sweep.

These early sloops were propelled by geared steam turbines; this was carried over to the 'Black Swan' Class sloops, and even to a fair proportion of the 'Algerine' Class. In this respect, they were ahead of all the wartime corvettes and nearly all of the frigates, as production of the turbine blades could not be maintained in wartime Britain.

The critical shortage of anti-submarine vessels in the early years of the war took all of these sloops away from minesweeping duties and they did not return, their big quarterdeck winches being replaced by depth-charge racks. So they helped develop the design of the large wartime sweeper fleet, but even the war-built 'Flower' Class corvettes carried out more operational sweeping than did these fine sloops.

'Halcyon' Class

The appearance of the first group of ships of this new class in 1928 was a great event for the minesweeping service. Twenty-one ships of the class were built, the last appearing just at the outbreak of war in 1939. They were specialised minesweeping sloops and, although many wartime modifications were made to them, especially in new sweeps and in anti-aircraft weapons, their design showed clearly that they were well planned.

Ships of the class were under construction at the same time as the general-purpose 'Grimsby' Class, and also in parallel with the first of the A/A and A/S sloops of the 'Bittern' and 'Egret' Classes. They were somewhat smaller than the general-purpose sloops, a feature to be followed in the 'Algerine' Class, while the A/A sloops were to be larger—the 'Black Swan' Class were 75 feet longer than the 'Algerines'.

Above Speedy, a *'Halcyon', after being mined forward off Malta in May 1943. She has a list to starboard, and is flying 'not under control' signals.*

Below *A dan-buoy being dropped from the British fleet sweeper* Bramble *in 1945. The buoy, with its flag flying, has just hit the water, and the concrete sinker, ready for slipping from the davit, is about to follow.*

The general lines of the previous sloops were maintained, but with concentration on a clear sweep deck aft. The solid bulwark was cut away as far forward as the searchlight platform. The later 'Algerine' Class reversed this trend, as it was found essential to have covered sweeping messdecks on the upper deck, where the crew could stand by comfortably in between handling the sweeps at the end of a leg in the sweeping pattern.

Wartime alterations to this class kept them up to date as far as possible with the wartime sweeper classes. The single forward 4-inch gun was retained, with a half shield fitted in some ships. Sponsons were built on to the bridge wings, to take 20 mm Oerlikons, which were also added aft above the sweep deck. Type 291 radar was fitted at the masthead quite early in the war, and the heavier Type 271 lantern was added on the bridge a little later. The tall mainmast came down very early being replaced by a light goalpost, to carry the radio aerials.

For sweeping magnetic mines, a large LL cable reel was fitted on the starboard side of the sweep deck, as there was no room for it amidships. The acoustic sweep was handled by a light derrick forward, as in the 'Algerine' Class. The depth-charge equipment was retained, but only as a few traps with two throwers.

This class was split into two groups—the first with reciprocating engines, the later ships with turbines. They were capable of 17 knots when running free, a knot more than the 'Algerine' Class, and the latter was to be regretted later, when speed in taking up position was found to be important in fleet sweeping operations.

Many of this class gave distinguished service in the early war years; a full flotilla was stationed in North Russia, for both sweeping and escort duties with

Passing an 'A' sweep between the 'Algerine' Class Bramble *and the 'Catherine' Class* Foam. *A good view of the roomy sweep-deck of an Algerine.*

Halcyon, *the name ship of her class, as she was in October 1942. She has a large LL cable reel mounted on the starboard side of her sweep deck, and twin Oerlikons have replaced the after 4-inch gun.*

the famous Russian convoys. *Bramble*, a leader of the class, was lost at the end of 1942 on this duty, being sunk by German destroyers in a night action. Her name was carried on in a new unit of 'Algerines', commissioned early in 1945. The class continued right through the war in front-line service, a full flotilla being present at the Normandy landings in 1944, and continuing through the bitter fighting in the southern part of the North Sea in 1945.

Moored contact mines

The design of these mines was progressed by both sides during World War 1 to a simple, efficient design, with few variations. Well tested in war service, it was against this device that nearly all sweeping efforts were at first directed.

The mine consisted of a spherical, hollow steel shell, containing buoyant material together with the explosive charge; the latter was typically about 600 lb of TNT or, in later war years, of Amatol, Minol, or other more powerful explosives. A moored magnetic mine was introduced by the Germans, early in the magnetic campaign; but it was not as successful as the ground mines and was not widely used. The mine was designed to explode when any one of about half a dozen horns, protruding outwards from the casing, was fractured by contact with a hard object, such as a ship's hull. These horns contained a chemical which, when released, activated the firing mechanism. The mine was moored to the sea bottom by a heavy iron or concrete sinker, from which a wire ran upwards to the underside of the casing. The depth below the surface at which the device would float was predetermined by the length of this wire. Sometimes the mine would have a delayed-action mechanism controlling the wire's release from the sinker, with the result that if the lay was suspected, a prompt clearance operation by minesweepers would not reveal the presence of the mines, which would rise to their calculated depth later.

Sweep obstructors, designed to part the wire sweeps before the mine's mooring could itself be cut, were introduced by the Germans in 1940. One type was an explosive conical float, some three feet long, and containing 1¾ lb of explosive; the destructor fired when the sweep wire reached the base of the float. Other types were static cutters, fixed to a larger float and, finally and most effectively, a heavy chain mooring on the mine which, by 1945, had made these devices very difficult to sweep, since even the heaviest sweep wires in use in the fleet sweepers could not drag the mines out of the swept channels.

Another device was the attachment of 'snag lines', kept on the surface by small floats, and attached to the mine's firing mechanism. These were especially dangerous for landing craft during assault operations.

Moored contact mines could be laid by any craft, but destroyers and motor torpedo boats, operating at night, were the most common means used by the Germans, especially in the North Sea and English Channel in the early war years. Indeed, there was a period when German E-boats were making fresh lays in the same or nearby positions every other night, when the Moon was not full. German raiders, operating around the world, also laid moored minefields off Australia, New Zealand, and South Africa. Japanese minelayers were also widely used.

One point of interest is the type of damage caused to the ship striking the mine. The contact mine would blow a large hole in the ship's hull, or even blow off the bow or the stern, the effects being immediate. A ground influence mine could have the same effect in shallow water, but equally it could break a ship's back, without making large holes in the hull. The result was just as dramatic, since the damage would at best take many months in dock to repair and, at worst, would cause the ship to be written off as a 'constructive total loss'—quite a common result with destroyers and frigates in the latter years of the war.

Moored mines were subject to two uncertainties in their eventual disposal. Some were timed to render themselves harmless by sinking to the bottom after a fixed length of time but, in a number of minefields, this mechanism did not function correctly and the mines remained floating beneath the surface. Others of the same type did not receive their sterilising mechanism because of the production difficulties in the last year or two of the war. This formed a special hazard for the postwar clearance sweepers, since the minefield charts were unreliable.

Probably the greatest uncertainty was the weather. Both in Europe and in the Pacific countless moored contact mines broke from their moorings in gales and drifted downwind, forming a hazard to any ships in the vicinity and, indeed, causing a significant number of ship casualties. Floating mines could be found far out into the North Atlantic and drifted right across the Pacific to the western coast of the United States. When the great mine barrier, off the East Coast of England, was cleared by flotillas of fleet sweepers after the war, there were great stretches of water where no mines remained at all. Even today mines are washed up on that coast during gales, or are recovered by fishing boats in their nets.

Wire sweeps

These were only used against floating mines. The gear used in the Royal and Commonwealth Navies and in the United States Navy was to all effects identical; the description which follows uses Royal Naval terms for convenience.

The sweep consisted of a flexible steel wire rope, 'laid up', or woven, in such a way that it was smooth for handling. It was stored in the sweeper on the drum of a special winch, located aft on the sweep deck. These winches were of the largest size it was possible to fit in each class of sweeper, in order to take the maximum length of wire, and the winches can be clearly seen in the photographs.

Control of the wire, through this winch, was critical to the success of the sweep. The Royal Navy preferred powerful and reliable steam winches, which could handle the sweep wire fast, while the United States Navy went in for equally

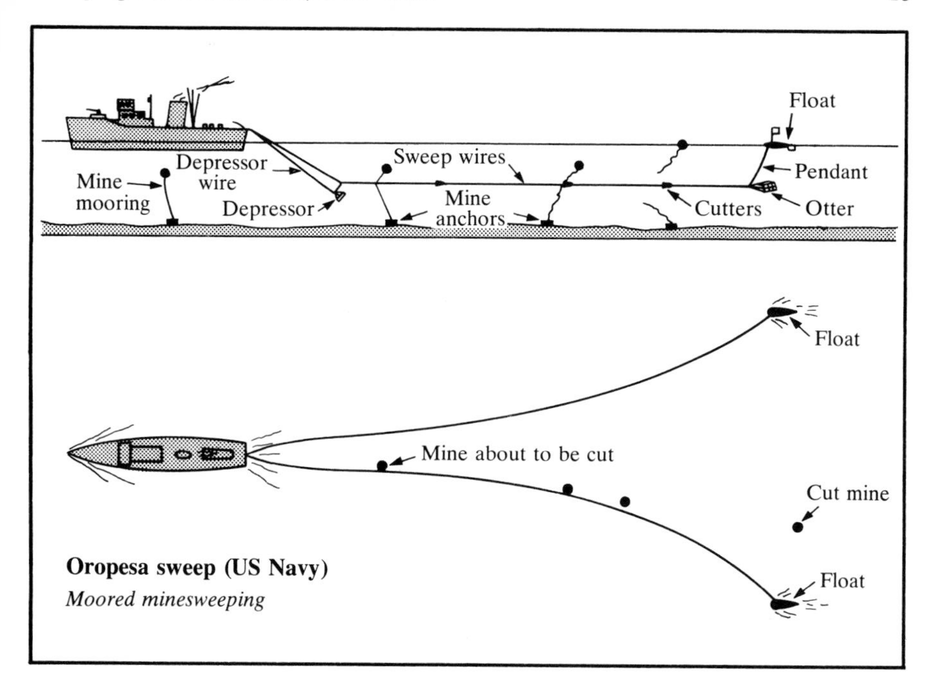

Oropesa sweep (US Navy)
Moored minesweeping

powerful electric winches. In Lend-Lease ships transferred to the Royal Navy, the latter did not find these as fast in operation as the steam winches. The wire was led out to the sea through heavy fairleads mounted right on the stern of the sweeper. The photograph of the stern of *Mutine* gives a good idea of this layout on the stern of a fleet sweeper.

In order to cut the mine's mooring wire as effectively as possible, explosive cutters were usually attached to the sweep wire near the deepest end (in an Oropesa sweep, at the float end, and in an 'A' sweep, on the bight of the wire). These had sharp cutters and small explosive charges, which went off when the mooring wire entered the teeth of the cutter.

Two types of sweep

The 'A' sweep used extensively in World War 1, was little used in World War 2, mainly because it limited the sweepers' freedom to manoeuvre. It consisted of a sweep wire, run out by one ship of a pair, and picked up by the second ship, brought in through its stern fairleads and secured to a quick-release slip. The sweepers usually operated in pairs with this sweep; more ships could be secured together with the sweeps between them, and in line abreast. But the more ships, the more clumsy the formation became, and they would all be operating in unswept and dangerous water.

The Oropesa sweep consisted of a single wire, which had a steel torpedo-shaped float attached to its outer end. The float was stowed right on the outer quarter of the sweeper, next to the minesweeping davit. It would be hoisted out by the davit, the wire and other gear attached, and then it would be dropped from the davit and the sweep wire paid out.

Two other items were needed to make the sweep wire travel through the water at the required depth, right along its length. One was the 'Otter', attached to the

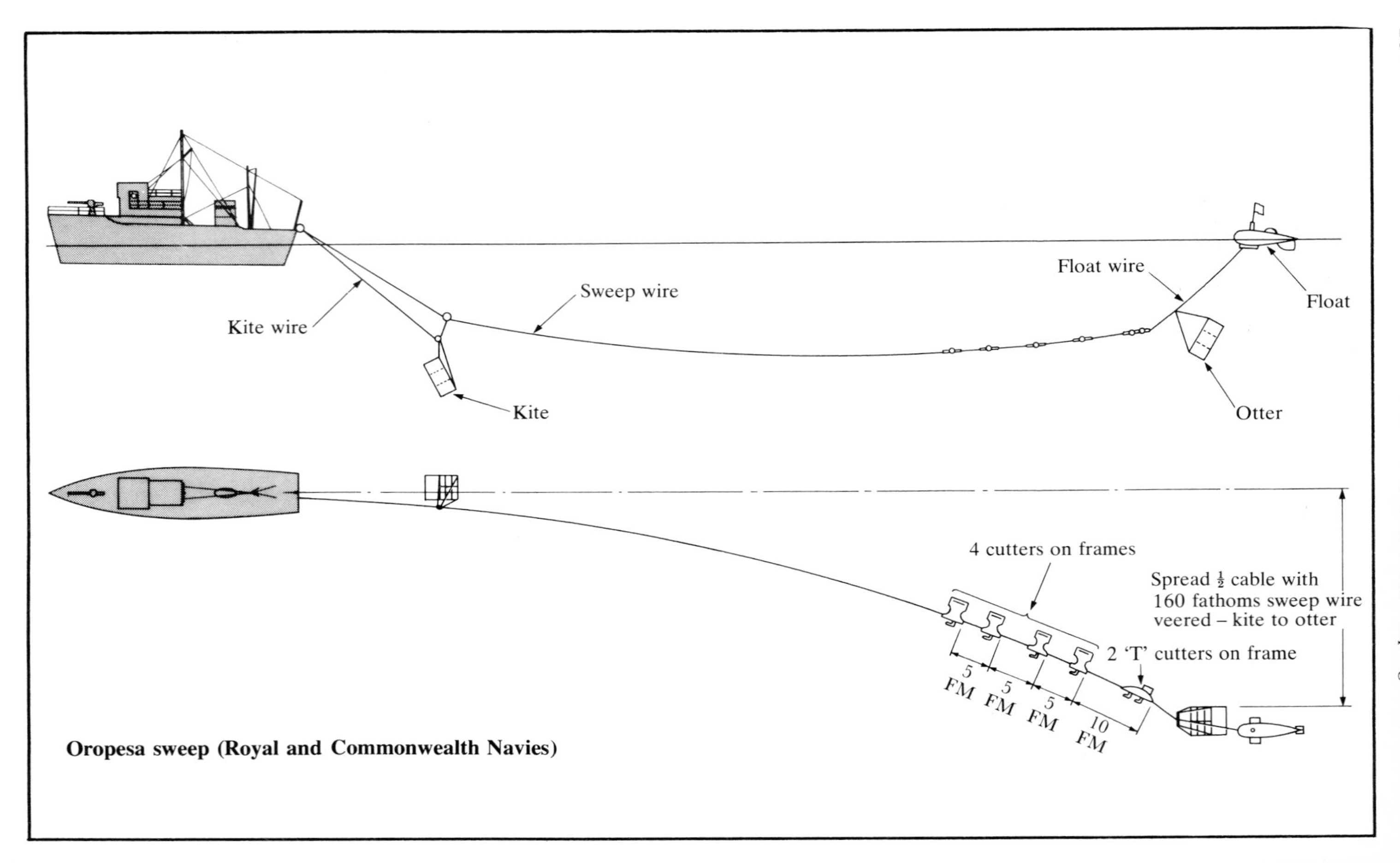

Oropesa sweep (Royal and Commonwealth Navies)

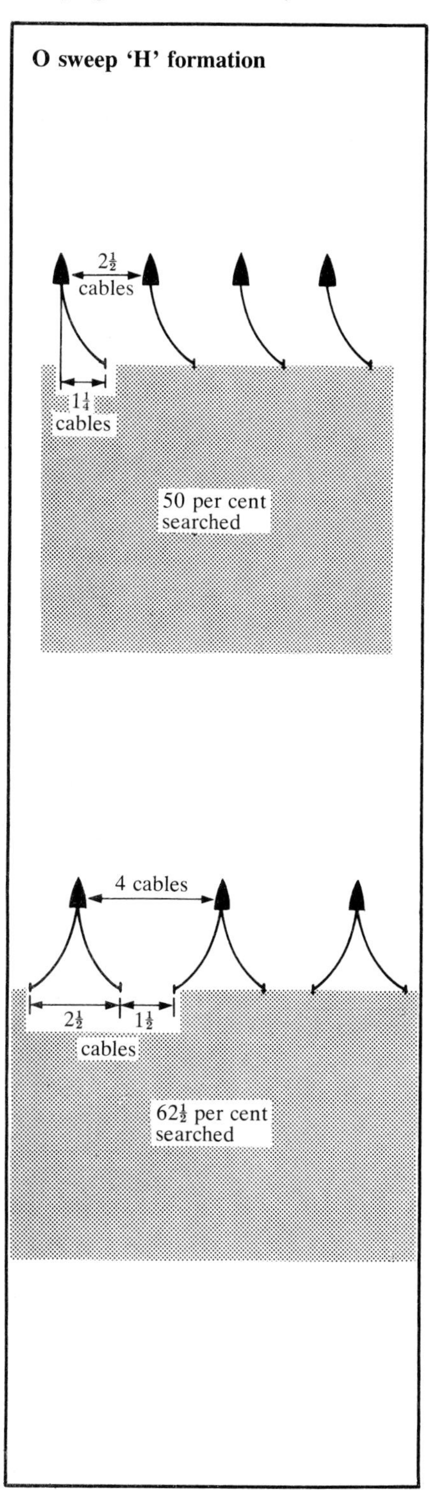
O sweep 'H' formation
2½ cables
1¼ cables
50 per cent searched
4 cables
2½
1½
cables
62½ per cent searched

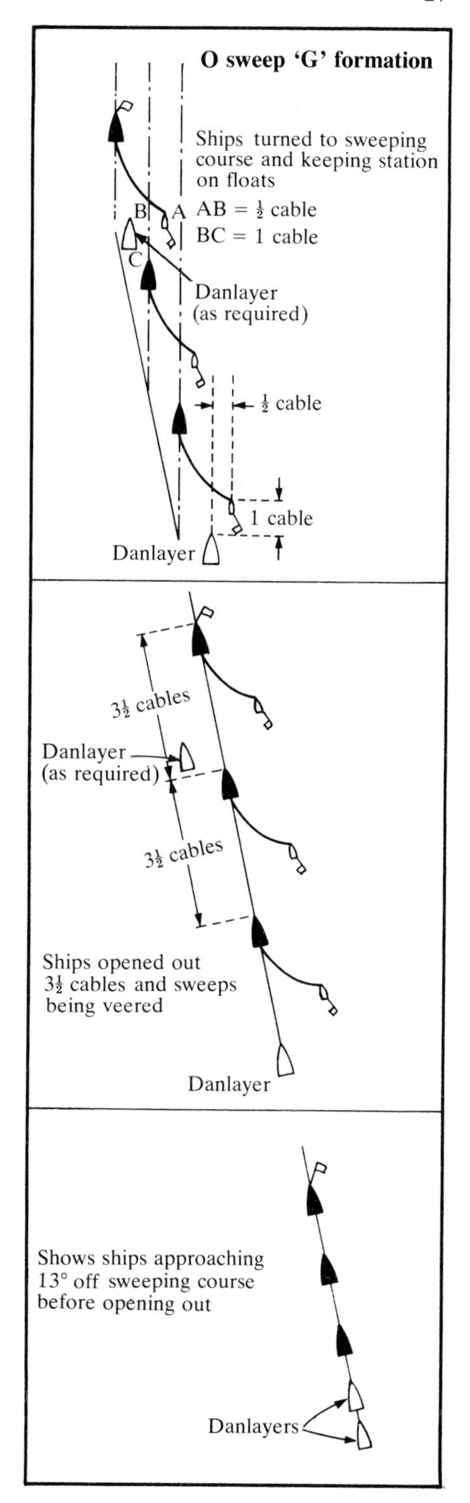
O sweep 'G' formation
Ships turned to sweeping course and keeping station on floats
B
A
C
AB = ½ cable
BC = 1 cable
Danlayer (as required)
½ cable
1 cable
Danlayer
3½ cables
Danlayer (as required)
3½ cables
Ships opened out 3½ cables and sweeps being veered
Danlayer
Shows ships approaching 13° off sweeping course before opening out
Danlayers

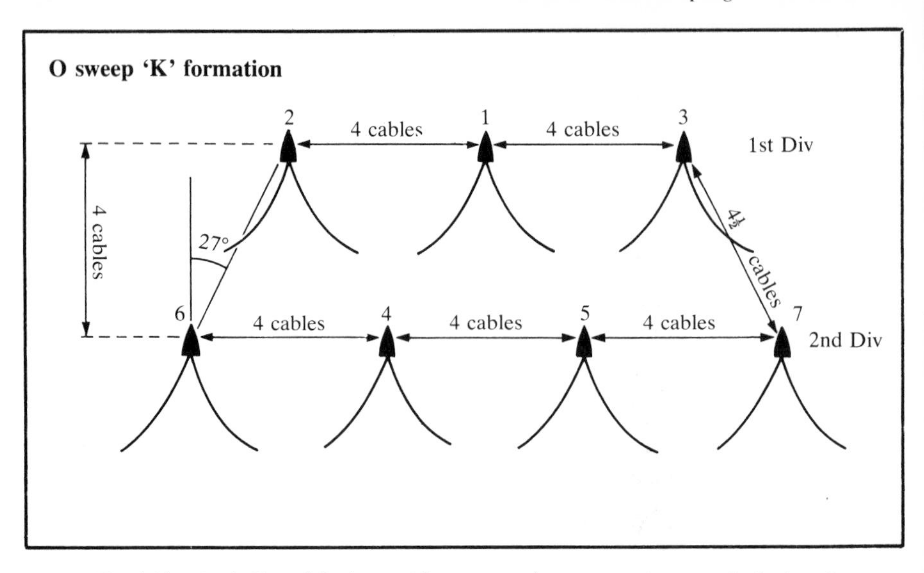

Below *The 'Algerine' Class* Mutine, *with a moored contact mine caught in her Oropesa sweep in the Adriatic in 1944. Otter and kite have been safely hauled in, but the mine is foul of the Oropesa float wire.*

Above right *ML 250, a Fairmile 112-foot type. She is fitted for Oropesa sweeping; the davits and floats can be seen on her stern.*

outer end and connected to the float by a wire, which equalled the required depth; the other was the 'kite', similarly lowered along the sweep wire from the stern of the sweeper, when the full length of the wire had been run out. These two were, in fact, identical and the only difference was the type of chain sling attached to them, to enable them to carry out their different functions. The 'kite/otter' consisted of a rectangle of flat steel, within which were several steel blades, running at an angle, so forcing it to run deep in the water. They were stowed on the quarters of the sweeper, in racks just inboard of the Oropesa floats.

When the sweep wire was running, the Oropesa float would be racing through the water; so, to keep it visible from the sweeper's deck, it had a vertical steel pole which carried a small flag, or sometimes a steel pellet painted red or green, attached to its head.

Sweepers could operate single (to one side of the ship) or double Oropesa sweeps; the single sweep was more commonly in use, and with the sweepers steaming in echelon, with each sweeper steaming with its bow just inside the float of the ship ahead. In this way all the ships, except the leading one (which usually turned out to be that of the Senior Officer!), were steaming in water which had already been swept. At the end of each lap of the sweep, the ship would wind in its wire to the 'short stay' position, to avoid fouling the sweeps of the other ships while turning to start the next lap.

The water swept would be marked by dan-buoys; these were developed from ordinary fishermen's marker buoys. They consisted of a long pole, with buoyant pellets attached half way down, from which the mooring wire extended to a concrete sinker. The buoy had a weight attached to its lower end, to keep it upright in the water, and brightly-coloured flags were flown from its top, to enable the sweepers to keep exactly to their lap course.

Accurate navigation was vital to good minesweeping; in flotilla leaders, and in some other ships, a machine known as Taut Wire Measuring Gear was installed. Piano wire was paid out astern of the sweeper, so that the distance to the minefield from the nearest headland could be accurately measured.

Chapter 2

Transition to secret weapons, 1939–1940

The European mine war started much as the Admiralty had predicted. However, within months the Germans had introduced the magnetic mine, and the minesweeping campaign in United Kingdom waters reached crisis proportions.

This was typical of the peacetime thinking of the period, which was too relaxed until 1938, then was followed by the astonishing adaptability of the Royal Navy, in meeting the new threat. Never again in this war were such dramatic developments in sweeps and tactics required as in the 16-month period which this chapter covers.

The war opens up

As soon as war was declared, the German campaign using moored contact mines started, as had been expected. German destroyers and torpedo boats laid fields off the English East Coast and in the Thames Estuary, by night. Three German seaplanes were even seen laying mines in daylight off the East Coast. The Germans were thought to have started the war with a stock of 200,000 moored mines.

The plans to increase the wire sweeping fleet, made before the war, were quickly put into action by the Admiralty. Some 800 commercial trawlers, drifters and whalers were requisitioned, fitted out with wire sweeping gear and their fishing crews quickly trained. Then they were back at sea, sweeping the war channels off the coasts. This was a big operation and was not accomplished without mishap. At Sheerness, a number of trawlers, just converted, moored to a pier for the first time. Everyone on board went ashore for a beer; but two tiers of trawlers had not been well secured. They came adrift and drove off on the tide up the river towards Chatham. Consternation all round, until the trawlers floated back with the tide at dawn and were secured to the pier without damage.

The available fleet sweepers were at once in great demand. Harwich, for example, had just six, with a dozen trawlers. Compare this with the 38 fleet sweepers, and MMSs and trawlers running into hundreds, which the port was to have by 1945.

The paddlers were to bear the brunt of the wire clearance sweeping in home waters during this period. Five flotillas of them were at sea very quickly, but one of the two flotillas of the new 'Halcyon' Class fleet sweepers was taken away for escort duties.

During the winter gales, large numbers of German contact mines were washed ashore on the East Coast—80 between Harwich and the Wash, another 100 north of there.

At the turn of the year, the Royal Navy sweeping forces, actually at sea on operations with trained crews, looked like this: *searching forces* consisted of 150 trawlers and 100 drifters: *clearing forces* were 16 fleet sweepers and 32 paddle sweepers. Apart from the growing fleet of requisitioned fishing vessels, 30 of the new 'Bangor' Class fleet sweepers were already on the stocks and to be in service by 1941; and nine of the new mine destructor ships were being fitted out.

However, the strain on the sweeping forces was already so great that nearly all the sweepers had to be concentrated on the East and South Coasts of England, with just a few left on the West Coast to keep the ports and important headlands clear. The searched channels around the West Coast had to be abandoned.

The magnetic mines appear

The first warning that something strange was happening came when ships started to be sunk or damaged without any contact mines being seen. Both merchant ships and warships were sunk, and fast-moving destroyers seemed particularly vulnerable. *Gipsy,* one of a force of five entering Harwich, was hit and grounded across the channel with a broken back, *Blanche* was sunk, and the cruiser-minelayer *Adventure* only just made port after being damaged.

Aircraft were seen laying mines by parachute at night, and big numbers of trawlers were sent to observe the laying positions. There were unexplained explosions around the coast, which were later discovered to have been magnetic mines going off prematurely. Casualties became so serious that at one stage the Thames at Southend was closed for 36 hours, and the Humber for two days.

The Admiralty had expected a magnetic mining campaign, but had no effective sweeps. It was hoped that countermining with depth charges would set the mines off, but this did not work. Then the destroyer *Wivern* tried steaming at full speed round the wreck of a ship sunk by a magnetic mine, to see if she could set another one off and use her high speed to get clear of the explosion. She could not obtain any results, which was just as well, though her crew had been reduced to a minimum for the experiment! Valiant efforts were made to recover a magnetic mine so that the settings being used could be established and effective counter-measures produced. But at sea, it was not easy; the drifter *Ray of Hope*, one of the special Mine Recovery Flotilla, was herself sunk by a mine which she had specially trawled up in the Thames Estuary.

Then a magnetic mine was discovered on the shore near Southend on November 22 and, by a very brave disarming operation, the mine was recovered and its secrets revealed at the Navy's mining school, HMS *Vernon*, at Portsmouth. Now that the polarity and sensitivity of the mine were known, the Admiralty quickly produced effective sweeps. The first attempt was with 'skids'—magnetic coils mounted on a small wooden barge, and towed by a tug. Soon, 176 had been constructed and they met with some success. Skids were used until the later LL sweeps were in full operation, and were only finally withdrawn from some ports in 1944.

The British flair for improvisation was coming through by this time; in February 1940 four skids were seen outside Harwich, being towed in line abreast by three different types of ship and led by the Commodore in his yacht. They even detonated some mines! These skids were towed by anything which was avail-

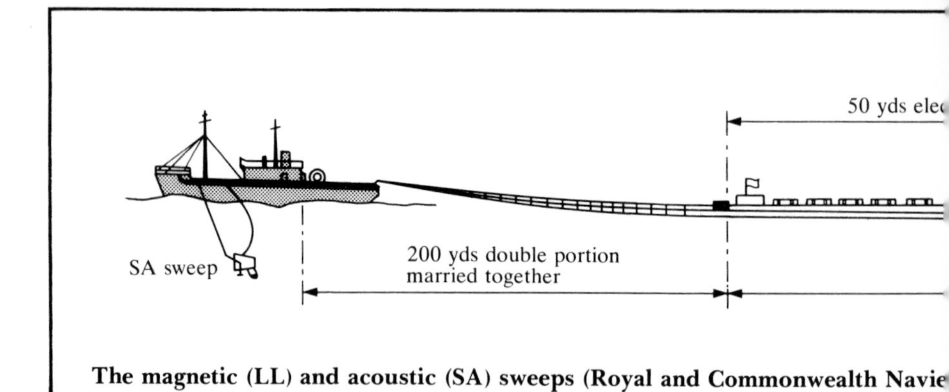

The magnetic (LL) and acoustic (SA) sweeps (Royal and Commonwealth Navie

able—one group at another port was towed by the drifters *Willing Boys, Playmates, Comfort,* and *Nautilus.* A high-speed version of the skid was also produced, and the destroyers *Skate* and *Scimitar* worked hard with this in the New Year.

Mine destructor ships and aircraft

These were both serious attempts at a solution, and they had some success. The aircraft were Wellington bombers, fitted with an enormous circular magnetic coil; they flew wingtip to wingtip in formations of three at wavetop height, up and down the buoyed war channels. They were called DWI aircraft sweepers, but today there seems to be no record of what those letters meant. Some of these aircraft got severely shaken up by near misses from the mines they had just lifted, and it was almost impossible to mark the cleared area with dan-buoys, so they did not continue their sweeping flights for long. They were transferred to Egypt, where they were quite successful in sweeping the Suez Canal.

The mine destructor ships numbered nine in all. The idea was to place a very large electro-magnet in their hold, three feet in diameter and 30 feet long. It was powered by two 300 kW diesel generators. An additional magnet was fitted over the bows, to increase the range. They were converted from coastal colliers or cargo vessels, and were chosen because their engines and bridge were aft, so keeping these vital parts as far as possible from the mine's explosion.

Borde, the first, started operations off the East Coast in January 1940; her first arrival at Sheerness created havoc, as her powerful magnet threw out the compasses of all other ships within miles. All the mine destructor ships were operating by the summer, and they had many successes in the early days—*Borde* sweeping as many as 20 magnetic mines in a single day. Her operations were fairly precarious; two Oropesa trawlers swept ahead of her, a drifter for life-saving followed close astern, and danlaying trawlers preceded and followed her. Most of the mines she exploded went off within 150 feet, some much closer, and the ship's company must have been in a constant state of shock. By late summer, the Germans were introducing pulse delay mechanisms into their magnetic mines, and the risk to the mine destructor ships then became unacceptable. In addition, one, *Queenworth*, was lost by air attack, while sweeping off the East Coast.

The 'Bosun's Nightmare' was the next try, used in parallel with these sweeps.

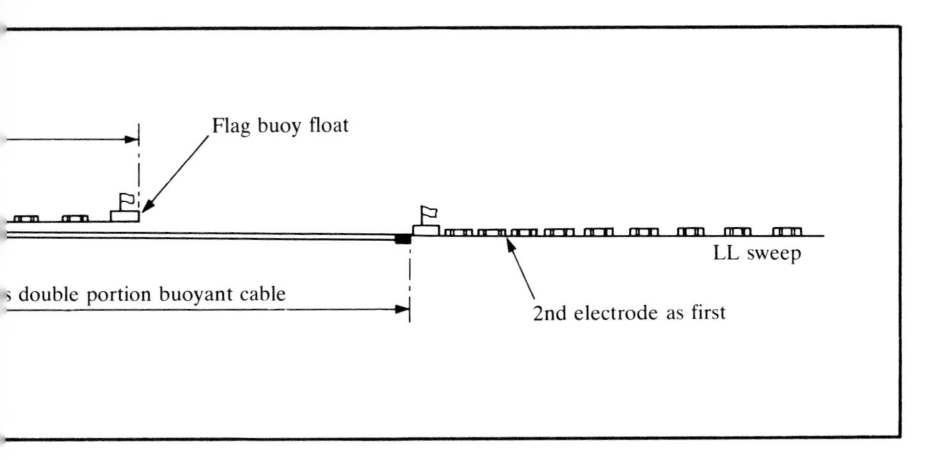

This was basically an 'A' sweep, a wire towed between two ships, but with 34 magnetic bars attached to it at intervals of 15 feet and on pendants each 40 feet long. The magnets trailed along the seabed. This contraption did lift some mines around the turn of the year. Before it was discarded, the number of magnets had been increased to 70 and the sweep wire was 700 yards long. No wonder the bosuns in the sweepers went crazy trying to disentangle it.

A similar 'Electric AA sweep' was tried, using armoured electric cable in place of the wire. Six trawlers were fitted with it, but the cable got damaged so frequently that it was soon abandoned.

The LL sweep appears

This was the final answer to the magnetic mine. It consisted of two long, buoyant electric cables, of unequal length, streamed from the stern of the sweeper. The latter had generators or batteries with which to energise the cables and, by pulsing the power on and off, a magnetic field was created which fired the mines.

Two small tugs, *Salvo* and *Servitor*, towed the first LL sweeps to sea early in 1940 and successes were soon rolling in—*Salvo* actually lifted her first mine on January 10. Then large numbers of trawlers were hastily fitted out with this sweep and, as the magnetic mines became more complicated to deal with, and the LL sweep proved itself equal to all of them, so the other methods were gradually phased out.

However, the long insulated LL cables were bulky and awkward to handle on board the converted fishing trawlers, and the steel hulls of these vessels made them vulnerable; so a large class of wooden minesweepers was placed under construction with the greatest urgency. It was these motor minesweepers which in fact kept the magnetic mine under control right through to 1945.

De-gaussing

A further and important refinement in this campaign was the de-magnetising of the steel hulls of ships, to reduce the risk of them detonating magnetic mines over which they might pass.

All steel hulls have a special north or south pole magnetic property, depending on where in the world they are built. It was found that by wrapping a big electric cable around the hull itself and by energising this cable through a generator, this property of the hull could be much reduced, in most cases to a relatively safe level.

This was called 'de-gaussing', after the scientist who had discovered the properties of magnetism. As soon as the method had been proved, all ships operating in the areas around the British Isles had these cables installed, with much success. In the early days, the cable was actually wrapped round the outside of the hull, just below upper deck level. In photographs of ships of that time, it stands out prominently—the first famous picture of a ship so fitted was that of the great Cunard liner *Queen Elizabeth,* arriving in New York on her first and secret voyage. Later the cable was fitted inside the hull, to avoid weather damage.

'Wiping' was a similar precaution—cables were dragged up and down the ship's hull, while a strong electric current was passed through them, and this served to give the ship a temporary measure of protection. Wiping stations were built at ports around the British Isles and, later the US Navy converted some YMSs and similar small warships, to wiping stations which could serve the fleet anchorages in the Pacific—these ships were designated YDG.

The mining campaign continues

While all this intensive activity to combat the new magnetic mine was going on, the moored contact mines were also still being laid by the Germans, and intensive wire sweeping operations had to be carried on without a break.

Before the end of 1940, two significant fields had been laid off the East Coast. One, off the Humber in December, closed the river for days and, of the 32 mines swept, most exploded on having their moorings cut. The other field, discovered off the Tyne also in December, was laid by German destroyers at night, and 12 fast sweepers hurried to the area to clear it. Nearly 50 mines were swept here and 18 ships were mined, including a minesweeping trawler.

Air attacks on the sweepers were a real hazard right through this period. There were losses, including the new fleet sweeper *Sphinx*, Senior Officer's ship of the

The fleet sweeper Sphinx *after taking a bomb on the fo'c's'le on the English East Coast in 1940. She later drifted ashore bottom up, after parting her tow in rough seas.*

Fifth Flotilla. She was bombed 15 miles north of Kinnaird Head, on the East Coast, and taken in tow, but the towrope parted in heavy seas and she capsized, drifting ashore bottom up. Much of the sweeping now had to be done by night, to avoid heavy sweeper losses.

During the war, the Admiralty issued to all sweepers a weekly summary of their operations, and in this period it makes stirring reading—of ships being mined, of the impromptu sweeps, and of the intensive and dedicated work of the sweepers.

They were doing well, too. By mid-May 1940, for example, the sweepers had lifted 345 moored mines and 167 magnetic—the latter figure increasing rapidly, week by week. Of those, the mine destructor ships had bagged 22, the DWI aircraft the same number, the skids 32, the magnet sweep 26, and the relatively new LL sweeps 47; and ten magnetic mines had been recovered ashore.

Dunkirk, and the German invasion

This was a particularly difficult period for the sweepers. The Germans were specially mining the East Coast naval ports during the Dunkirk evacuation; and the sweepers were badly needed off the beaches as well. Within a few days one new fleet sweeper, four paddlers and two trawlers were sunk, and five other fast sweepers damaged. Then, immediately after the evacuation had been miraculously achieved, the sweepers joined the anti-invasion patrols round the coast. The LL sweepers had to keep going by night, too, as it was expected that the Germans would sow magnetic mines around all the naval ports on the night before the invasion. German aircraft were minelaying in all the French ports and British LL sweepers went over to help keep them clear during the final stages of the evacuation.

The long, hot summer

All through that summer of 1940, and right through the Battle of Britain, the sweepers ploughed on. The numbers of LL sweepers in service were growing fast, and the bag of mines swept was growing with them.

By June, 70 LL trawlers were in commission and, by October, there were 250. By October's end the bag had risen to 940 magnetic mines, and 640 moored contact, showing how the battle had changed its emphasis since the early days. Since the search only a few months earlier for the first magnetic mine to fall on shore, over 200 had now been dropped on the ground around London alone.

Some destroyers had bone-shaking experiences with the magnetic mines. In September, three of them steaming fast in line ahead, lifted two mines near the Tongue lightship in the Thames Estuary. Two more, off the Humber in October, went up just astern of *Intrepid* and *Icarus*, as they passed by at 25 knots.

A stormy winter

The moored contact mines continued to appear, in enough numbers to require the wire sweepers to search all the swept channels continuously. In October German destroyers, now based in occupied France, laid a big field off Falmouth, which gave a lot of trouble. The field stretched from the Lizard to the Eddystones, and Falmouth itself had to be closed. A large number of sweep obstructors and delayed-action sinkers had been included in the field, together with a new type of explosive conical float which fired when the sweep wire touched its base. Another, larger float also made its appearance here and, though it was not explosive, it carried much larger cutters, doing a lot of damage to the sweeps.

In the last two months of the year, E-boats, by now also based in France, laid fresh fields every other night off Dover. Again, many obstructors to the sweep wires were included. The area was also full of wrecks, so that the sweepers had a hard time clearing each field as it was laid. At the southern end of their sweeping legs, they were within range of the guns at Cap Gris Nez, and were subjected to air attacks at frequent intervals.

The year finished on an ominous note, with the newest secret weapon, the acoustic mine, beginning to make its presence felt.

Last, but by no means least, the bag of mines swept in the Nore Command had by the end of the year gone up to 1,214 magnetic, 767 moored contact, and 105 of the new acoustics.

Minesweepers around Australia and New Zealand

In June 1940, mines were laid off Auckland, in New Zealand, by a German raider, the *Orion*, and the liner *Niagara* was sunk. Three Australian sweepers went to clear the field and an intensive search sweep was made in the Bass Strait, where loud offshore explosions had been reported by lighthouse keepers.

By the end of June, 12 minesweepers in five groups were operating out of Sydney, and this rapidly increased to 70 ships operating from six ports around Australia. Routine search sweeps were maintained in all swept channels, and just as well, as another German raider, the *Pinguin*, laid fields off Adelaide and Hobart in October. The raider escaped, but 12 mines were accounted for, while two ships were sunk.

In December, two further fields laid by this same raider were discovered off New South Wales. The sweepers were under great strain, hurrying from one trouble spot to another over great distances. In spite of their efforts, two more ships were lost on these mines.

Minesweepers off South Africa

The German raider *Atlantis* laid a field of 92 moored contact mines early in May 1940, off Cape Agulhas, some 175 miles from Cape Town. Some of these exploded spontaneously in the heavy sweeps, so alerting the minesweeping forces to their existence. Sweepers were in short supply at that time, but a clearance flotilla of four ships was quickly on its way, arriving in the area on May 14. They searched all day with an 'A' sweep without result, but next day two mines were discovered and the flotilla was recalled to Cape Town while further plans for clearance were made.

A force of six ships was organised, two of which acted as danlayers. At the end of the month and, in early June, these ships swept the area many times, but only 11 mines were found. Many of the mines must have broken adrift in the strong seas, and some of them may have been fitted with sterilisers.

'Bathurst' Class Fleet Minesweepers (Royal Australian Navy)

This was a unique class in its own right, consisting of 60 ships built in Australia, and was the only class of war-built fleet minesweepers designed and built outside the United Kingdom and the United States.

The class was in the design stage in Australia in 1939, and the ships were first rated as AMSs (Australian minesweepers), but were later widely referred to as

Bathurst, *the lead-ship of her class, designed and built in Australia. She has full Oropesa sweeping gear aft, but little A/S equipment.*

corvettes. In the early war years it was thought that mine warfare, rather than U-boats, would be the main threat in Australian waters; early ships of the class reflect this emphasis, while some later units were completed with a full A/S configuration.

There are resemblances in the 'Bathurst' Class to both the Royal Navy's 'Flower' and 'Bangor' Classes. Their full speed running free was originally 15 knots, later increased to 16, and their sweeping speed was 10 knots. The Admiralty comment on the class was that it was a bit larger than the 'Bangor' Class, and rather better fitted.

In late 1939, four ships of the class were laid down for the RAN and, in January 1940, the Admiralty placed an order for ten, for delivery at a rate of one per month in 1941. Further orders were placed to a total of 36 by the RAN, ten more were ordered by the Admiralty, and four by the Royal Indian Navy.

Of the 20 ships ordered by the Admiralty, all were manned by the RAN on commissioning, and 13 of these remained under the operational control of the Admiralty. These served in the Indian Ocean and the Pacific, while some were used in the Mediterranean and took part in the invasion of Sicily. All ships of the class were in service by mid-1944, with most of them being completed a year earlier than that. They were handy little ships and performed very well in a variety of roles, not only protecting Australian waters but also participating in the earlier American assaults on islands near Australia, and providing the initial fleet sweeper cover for the British Pacific Fleet in 1944–45.

Requisitioned trawlers, drifters and whalers (RN)

Of a total of 745 ships 402 were trawlers, 287 were drifters and 56 were whalers. These figures have never been published before in this form and are quite astonishing. They cover only those vessels used for minesweeping; at least as many again were requisitioned for other duties—anti-submarine patrol and escort, armed patrol work, carrying barrage balloons, boom defence, carriage of oil and stores, torpedo recovery, radar instruction, submarine tenders, target-towing, examination vessels, and harbour service of all kinds.

The figures themselves are approximate only—quite a number of vessels were used for more than one type of service. It is, however, noteworthy that trawlers and drifters were not often switched from anti-submarine work to minesweeping, and vice versa. This is a different picture from that of the fleet sweepers and is due to restricted space, for depth charge or sweeping equipment aft, in these smaller vessels. In war-built Admiralty trawlers, both types of equipment were installed in some classes, where the space could be specially designed to take them.

Looking at these requisitioned vessels together with the war-designed trawlers, two trends stand out. First, the rate of war losses among the requisitioned vessels was comparatively high, and secondly, comparatively few Admiralty trawlers were built during the war, due to the big supply of commercial vessels.

Trawlers

Of the 402 used for sweeping, a large proportion were elderly; 111 were completed before 1914, 134 during World War 1, 131 between 1920 and 1930, and only 26 in the years 1931–39. Two were actually completed before 1900, but were returned to commercial use early in the war!

Despite their age, these ships served all over the world, far from their native fishing grounds. Used at first for wire sweeping against moored mines, many were converted to LL/SA sweeping against ground mines—space precluded their being used for both types of sweeping.

As the war years passed, the new motor minesweepers, coming into service in great numbers, largely took over from the trawlers the influence sweeping in home waters, but the prewar trawlers were still active to the war's end. Quite a large number switched from one type of duty to another. Thirty-eight served in other roles before they became sweepers—all except one as armed patrol vessels. Fifty-two, on the other hand, switched from sweeping to other less exacting roles, as the MMSs came into service; 24 to carrying oil fuel, seven to carrying stores, 15 reverting to armed patrol, and six to barrage balloon and boom defence. No less than 62 were returned to their commercial owners before the war's end—22 in 1944, the rest before VE-day in 1945.

Losses were high—a total of 84 vessels; from three in 1939, yearly losses rose to 33 in 1940, and 23 in 1941; then they dropped, to nine in 1942, the same in 1943, two in 1944, and five in 1945. The importance of these ships in 1940 and 1941, especially in European waters, comes through very clearly.

Drifters

Here also elderly ships were put to good use. In World War 1, four main classes of drifters were ordered by the Admiralty in large numbers; they were known as Admiralty drifters, and as the 'Castle', 'Strath', and 'Mersey' Classes.

Many were cancelled, but many more were sold in 1918 for commercial use. These were requisitioned again in large numbers in 1939, and of the total of 287 drifters used for sweeping duties, 163 were of these four classes.

War losses, again, were comparatively high—a total of 55, equally divided between the ex-Admiralty drifters and commercial designs. Their great value, like the trawlers, was in 1940 and 1941, when they carried a very heavy load. From two lost in 1939, losses rose to 20 in 1940, 19 in 1941, then dropped dramatically to four in 1942, six in 1943 and two each in 1944 and 1945.

As many as 50 of these drifters were diverted to other roles, as the MMSs came into service, and 42 were returned to their commercial owners before the end of the war. But they provided faithful service right through to VE-day; one drifter, a stores carrier, was sunk off the Normandy coast in 1944, some 40 years after she was built!

Whalers

The large number of whalers requisitioned may seem surprising, 56 for the RN, but this reflects the importance of the whaling business in prewar days.

These were sturdy little vessels, well used to heavy seas, but their forward deck (normally used for handling whales) was very low, and so was usually awash, needing a raised gangway from the bridge to the forecastle.

War losses, at eight, were quite high; seven more returned to commercial use before VE-day. They were comparatively new, compared with the trawlers—of those used for sweeping, 41 were built in the inter-war years, 1920–30, and 15 in the eight years leading up to the war.

Minesweeping trawlers in World War 2

	Naval trawlers			Requisitioned						
Country	**Bought in**	**War built**	**Total**	**Trawlers**	**Drifters**	**Whalers**	**Total**	**Grand Totals**	**Losses No**	**%**
United Kingdom										
M/S	64	20	84	402	287	56	745	829	186	22½
A/S–M/S		189	189				*	189	37	20
Totals	64	209	273	402	287	56	745	1,018	223	20
Commonwealth										
Australia				11			11	11	2	9
Canada	3	12†	15	10			10	25	1	4
India		22†	22	1			1	23	0	—
—										
New Zealand		17†	17	8			8	25	1	4
South Africa				10		35	45	45	3	7
Totals	3	51	54	40		35	75	129	7	5½
Allied countries										
Belgium				4			4	4	0	—
France				34			34	34	5	15
Holland				23			23	23	0	—
Norway				10		30	40	40	6	15
Totals				71		30	101	101	11	11
United States				18	42		60	60	0	—
Totals				18	42		60	60	0	—
Combined totals										
All navies	67	260	327	531	329	121	981	1,308	238	19½
United Kingdom & Commonwealth	67	260	327	442	287	91	820	1,147	227	20

* Many others in the Royal Navy were requisitioned for anti-submarine work; but, unlike the naval trawlers, which were specially designed, they were not fitted for both duties.

† These ships were fitted for the dual role, A/S–M/S. Some probably concentrated on A/S work, but most would have performed M/S work from time to time.

The United Kingdom numbers of requisitioned vessels underline not only the scale of the mine warfare campaigns in northern European waters, but also the heavy dependence on these vessels of the Royal Navy, until the war-built MMSs and BYMSs were in commission.

Magnetic mines and sweeps

The principle of the magnetic mine was by no means new. The British had started laying the 'M-sinker' in 1917; but development had been frozen between the wars, as in other branches of naval warfare, and not until early in 1939, did the Royal Navy reconsider this mine. In the interval, the Germans had developed it, and were ready to launch a magnetic mining campaign.

In World War 2, magnetic mine warfare can be broken down into three periods:

1939–40 First German mines found, and LL sweep developed.
1940–42 Combination with acoustic mines, and arming and period delays, increasing the sweeping requirement many times.
1942–45 No change in German technical effort. Main research concentrating on how to sweep British magnetic mines after the war.

The early sweeps showed good initiative and emergency action; but the ultimate in sweeps turned out to be the LL, especially when fitted in large numbers in the new wooden MMSs. Fleet minesweepers did little influence-sweeping before 1944, and the American-built YMSs, excellent ships, appeared in 1943—by that time the main magnetic sweeping developments had already occurred and, whilst energetic and widespread sweeping was required, the main threat to Allied shipping from purely magnetic mines had been largely overcome.

The magnetic mine

The principle was a simple one. It incorporated an electrical unit, which picked up the magnetic field of a ship passing over or close to it. When the strength of the field was strong enough to close the contacts of this unit within the mine, it would explode.

The mine could be set at various levels of sensitivity—these were called 'coarse' or 'fine'—so varying the strength of the magnetic field required to fire it. The magnetic mine was almost invariably laid on the ground; it looked like a long, cylindrical tube and, internally, the magnetic unit with its batteries would be located at one end and the explosive charge at the other.

It could be laid easily by aircraft, using a parachute, and this became the usual method; but German E-boats, operating at night, also laid this mine in large numbers. It was ideal for use in shallow waters, such as the southern North Sea and the English Channel, and the damage it caused to a ship passing overhead was different from that of the contact mine. It would blow a hole in the ship large enough to sink her, if the explosion was close enough. An equally destructive alternative was that the explosion on the seabed would break the back of the ship and dislodge even her main engines from their seatings, let alone damaging all the electronics. So many ships were written off because the reconstruction work, compared with the cost of a new ship, would not be worth while.

In the later stages of the magnetic mining campaign, when the Royal Navy's counter-measures were becoming effective, the Germans introduced a number of refinements to the firing mechanism, which carried over to the later acoustic and pressure mines. These refinements consisted of an arming delay, which meant that the magnetic firing mechanism did not become active until a set number of hours or days after the mine had been laid; and a 'pulse delay mechanism', which meant that a sweeper could pass over the mine and activate the mechanism a set number of times before the mine would actually explode.

These new additions meant that even when the sweepers knew that a mine was there, they needed to sweep the area many times before it exploded, or could be assumed to be dead.

Two of the German magnetic mine types should be mentioned here:

The Type G mine

Introduced in May 1941, and known to the Germans as M.101. This one could be laid without a parachute and was therefore very difficult to spot when laid at night. It also had a much coarser sensitivity, which needed a magnetic pulse lasting up to ten seconds, before it would fire; so it could catch a large, slow merchant ship passing over, but the sweepers ahead of her would miss it.

The 'Sammy', or Type AM Mark 1

Introduced in the autumn of 1941, and known to the Germans as MA.1. This was a combined magnetic/acoustic mine, so that a magnetic impulse would in turn activate the acoustic mechanism. The magnetic side was set fine, and the sweepers then had to cover a much larger area of seabed, and with a complicated series of sweeps. All the safety precautions for the sweepers had to be revised when this mine appeared, and this was an example of the constant battle of technology between the minelayers and the sweepers, even when the principle of dealing with the magnetic mine had been mastered.

Early magnetic sweeps

The magnet sweep (Mark II*)

This was an 'A' sweep with 34 magnetic bars, spaced 15 feet apart, attached to the steel sweep wire and on wire pendants each 40 feet long. This was later increased to 70 magnets, each 27 inches long, spaced ten feet apart, on a sweep wire 700 feet long. Elliptical floats along the sweep wire kept the magnets about five feet off the seabed and the sweep was used in water up to 25 fathoms deep. It was at sea in December 1939, and was called the 'bosun's nightmare', due to the difficulty in handling it.

The electric 'AA' sweep

This sweep, which had existed in embryo since 1919, was used in October 1939 before the first German magnetic mine had been recovered and dissected. It consisted of two armoured electric cables, towed between the sterns of sweepers like an 'A' sweep. The ships were 1¼ cables apart and the sweepers activated an electric pulse in the cable of 300 amps. Six trawlers were fitted out to use this sweep but the cables had to move near the seabed to be effective. They were constantly being damaged, and thus this sweep, too, was phased out.

Electric 'skid' sweeps

These consisted quite simply of a magnetic coil, mounted on a wooden barge, or skid, and towed by any type of vessel. The coil was activated by batteries or by a generator in the towing ship, which would be 'de-gaussed'.

The first skid unit went to sea in December 1939, and soon 23 trawlers and 22 drifters were being fitted to operate it. The coil consisted of 200 turns of copper cable in one coil of 18 feet diameter, activated by a 20 kW generator. It was mounted horizontally, and the path it swept was estimated at 25 yards in ten fathoms. Its main disadvantage was that each mine lifted also destroyed the skid that did it.

Mine destructor ships

The coastal colliers converted were of 2,000 to 3,000 tons gross. A large magnet was placed in the hold, forward of the bridge; it weighed 450 tons, was 105 feet long, and had a core of 5½ feet in diameter. Two 300 kW diesel generators provided the main power for it, at 1,200 amps and 220 volts, the power to the magnet running at 4,000 amps. An additional magnet was mounted on a platform over the bow, to increase the range, and this was connected in parallel with the main magnet.

These ships were withdrawn from this role by the summer of 1940 when the introduction of pulse delay mechanisms placed them at great risk. They were then converted to single or double LL sweepers, with two standard trawler tails mounted on reels on either side of the after superstructure. Some of them operated in this role right through to 1945, based at Swansea; but others were converted to minesweeper maintenance ships, for use in the Mediterranean and the Far East.

Statistical sweeping

This deserves a mention, as it had a significant effect on the sweeping of influence mines from 1940 right through to 1945. Large numbers of sweepers were at sea at night, to plot carefully the positions in which aircraft were seen to lay mines by parachute. This reduced greatly the area of search for the sweepers

The mine destructor ship Springdale *in 1941, showing the additional magnet mounted over her bows. She has two 12-pr guns mounted forward of the bridge.*

Two short MMSs of the Royal Navy exploding a magnetic mine between their LL tails.

the next day, especially when arming delays and pulse delay mechanisms prevented the mines from being revealed by a normal searching sweep.

Not only was this method used effectively in the southern part of the North Sea in 1940–41, it was a major commitment for all the BYMSs and MMSs at night in the Normandy assault area in 1944; and in the last six months of the European war, a large land-based minewatching force covered the vital River Scheldt, where the Germans were constantly laying ground mines by night.

The LL sweep

This quickly became the standard basic sweep against magnetic mines from 1940 onwards, for all the Allied navies. It consisted of a pair of electric cables towed astern of the sweeper, and energised from on board. But this simple principle, to create a magnetic field which would fire the mine, needed much development. The field from a single straight wire carrying electric current was well known, but the effect of the return current in the salt water was not. Development trials were carried out under great pressure in late 1939, once the first mine was recovered.

The effective sweep resulting from these trials was a pair of electric cables, one long, one short, 'married' or lashed together. In the first operational LL sweep, the Mark I*, which went into service in March 1940, the long leg was 525 yards long and the short one was 125 yards long. Each leg terminated in a 50-foot electrode and the two cables were lashed together for the full length of the shorter cable, to cancel out the magnetic fields created near the sweeper herself.

The electric power was at first provided by banks of ordinary car batteries in the sweeper but, as the current was pulsed, these batteries became overheated. Therefore, diesel generators were installed in all LL sweepers, as soon as supplies permitted. A 35 kW generator was sufficient to energise the early LL sweeps, but the later Mark III, fitted in the MMS, needed a 108 kW generator. This gave double pulsing at 3,000 amps at 7 knots, the maximum continuous operating speed of the MMS. The 'Algerine' Class fleet sweepers were fitted with the Mark VI sweep, including a 220 kW generator, which gave double pulsing at 12 knots; but fleet sweepers were hardly ever used for LL sweeping before 1944, and then the gear gave quite a bit of unforeseen trouble.

The electric current was pulsed through the cables with either one or both polarities and it was found that, using a 60-second pulsing period, with the ships steaming at 8 knots, a swept path 270 yards long was achieved with each pulse.

An early refinement was for these sweepers to operate in pairs, or a combination of a number of pairs. They then needed to coordinate their pulsing carefully and electric lights, facing outwards abeam, were installed forward and aft in each sweeper. With these lights flashing to indicate the polarity of each pulse, the sweepers could ensure that they were pulsing together. The pulses, and the polarity of the mines, were called red and blue. In photographs of the period these lights can easily be identified, since black squares were painted round them to make them show up more clearly in daylight.

The production of the buoyant cables gave an opportunity for more improvisation. No such cable had ever been produced before; so the British manufacturers made an armoured, rubber-covered cable, which was kept afloat either by a large number of wooden pit props, or by equally large numbers of tennis balls housed in cylinders. Later, a real buoyant cable was produced, which needed no additional support.

This cable was bulky and needed careful handling, to avoid damaging it. So the MMSs, and later the BYMSs, carried a large reel aft on which the cable was stowed and large wooden roller fairleads were fitted on the stern to cut down chafe as the cable was run in or out. The batteries or the generators were housed in a room below the sweep deck. In the larger fleet sweepers, the 'Algerine', 'Raven', and 'Admirable' Classes, the cable reel was stowed just forward of the big minesweeping winch and light gun platforms were built over it in some classes. In others, notably the 'Bangor' Class, room on deck did not permit the use of a big reel and the cable had to be stowed on the deck below, increasing the risk of chafe.

Technical advances were made in the LL equipment during the later war years, but the principles remained the same. Slave control gear was introduced to mechanise the synchronisation of the electric pulses. Double-pulsing and larger capacity generators made their appearance, and there were special variations for use in large fresh-water rivers (this was called the LAA sweep) and in very shallow water using special wooden barges called ZZ craft.

Types of LL sweep in the Royal Navy

Class	Mark of sweep	Length of tails Long	Length of tails Short	Max sweeping speed	Power in kW	Pulse length	Cycle seconds
Tugs	I	525	125	8	70	5	60
Trawlers	II	525	125	6/8	70	5	65
MMSs	III	525	200	7	108	4½	60
'Bathurst' 'Bangor' 'Algerine'	V	525	200	11	42	5	40
Some 'Bangor'	VI	575	225	11/12	160	5	40
BYMSs	VII	525	200	10	360	5½	40
BAM ('Raven')	VII	575	225	12	710	5½	36
M/S ML	VIII	500	125	7	35	4½	60

Type LA and LAA could be fitted in MMSs, BYMSs, and some MFVs.

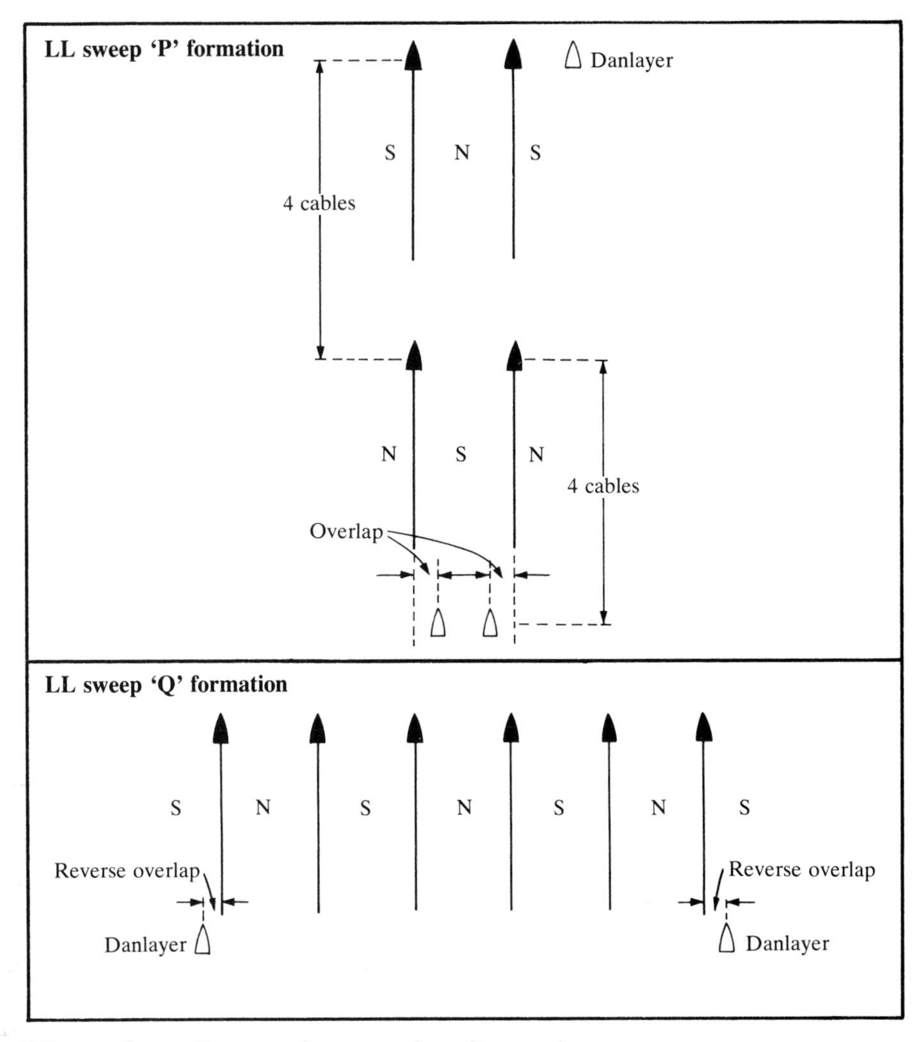

Magnetic and acoustic sweeping formations

Five basic formations were devised for ships sweeping against influence mines. These formations were first developed for the LL sweepers against magnetic mines. When the acoustic mines appeared about a year later the same formations were used when MMSs or BYMSs, for example, were sweeping with both LL and SA sweeps. Of these five formations, 'P' and 'Q' were the standard ones normally used on operations, 'P' for clearance, and 'Q' for search.

'P' formation

This was the normal clearance sweep for two or more ships. It covered a limited front but, if four or six ships were available, it dealt with any pulse delay mechanisms. All pairs swept with single pulsing, in the same polarity. Ships swept two abreast, four cables between pairs in column. Danlayers were stationed four cables astern of the rearmost sweepers, with a good safety overlap inside the LL tails.

'Q' formation
This was a high-percentage searching sweep, for mines of both polarities. It was used by two or more ships. The sweep covered a wide front, but did not deal with pulse delay mechanisms. Ships swept two or three in line abreast, but with not more than three ships in each sub-division.

The sweepers maintained 2½ cables station-keeping distance, using alternate polarities in synchronisation. The danlayers were stationed just outside the outer LL tails in this formation.

'R' formation
This was used in sweeping ahead of a convoy, or in clearing a marked channel. If a 100 per cent sweep for mines of both polarities was needed, then it was necessary to cover the area twice in single or double pulsing. The ships swept in pairs, in arrow-head formation, with the danlayers stationed just inside the outer LL tails of the last ships. Station-keeping distance was four cables from ship to ship, in column.

'S' formation
This was good for a clearance sweep, when 100 per cent cover for mines of both polarities was required. Three sub-divisions, or pairs, were the maximum number which could be used and the area needed to be covered twice with single pulsing. The pairs were disposed in quarter line to port, if the minefield lay on their port hand, and similarly to starboard, if the field lay there. The danlayers were stationed astern of the rear sub-division, and station-keeping distance was four cables in column.

'U' formation
This was intended for use against super-sensitive mines, even if they were mixed with more coarsely-set mines. Only four ships could be employed, with four cables station-keeping distance. The area needed to be covered twice if single pulsing was used, and the danlayers were stationed just inside the tails of the two leading ships.

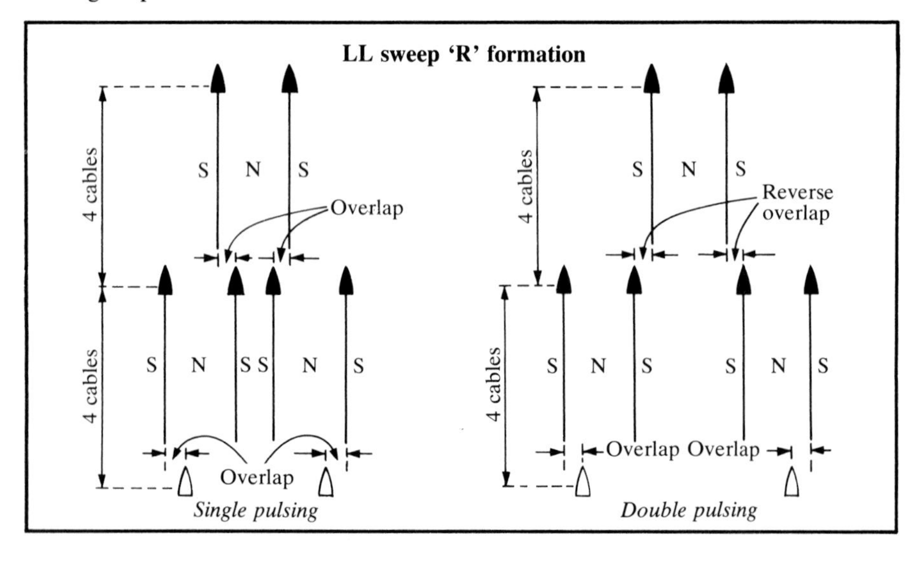

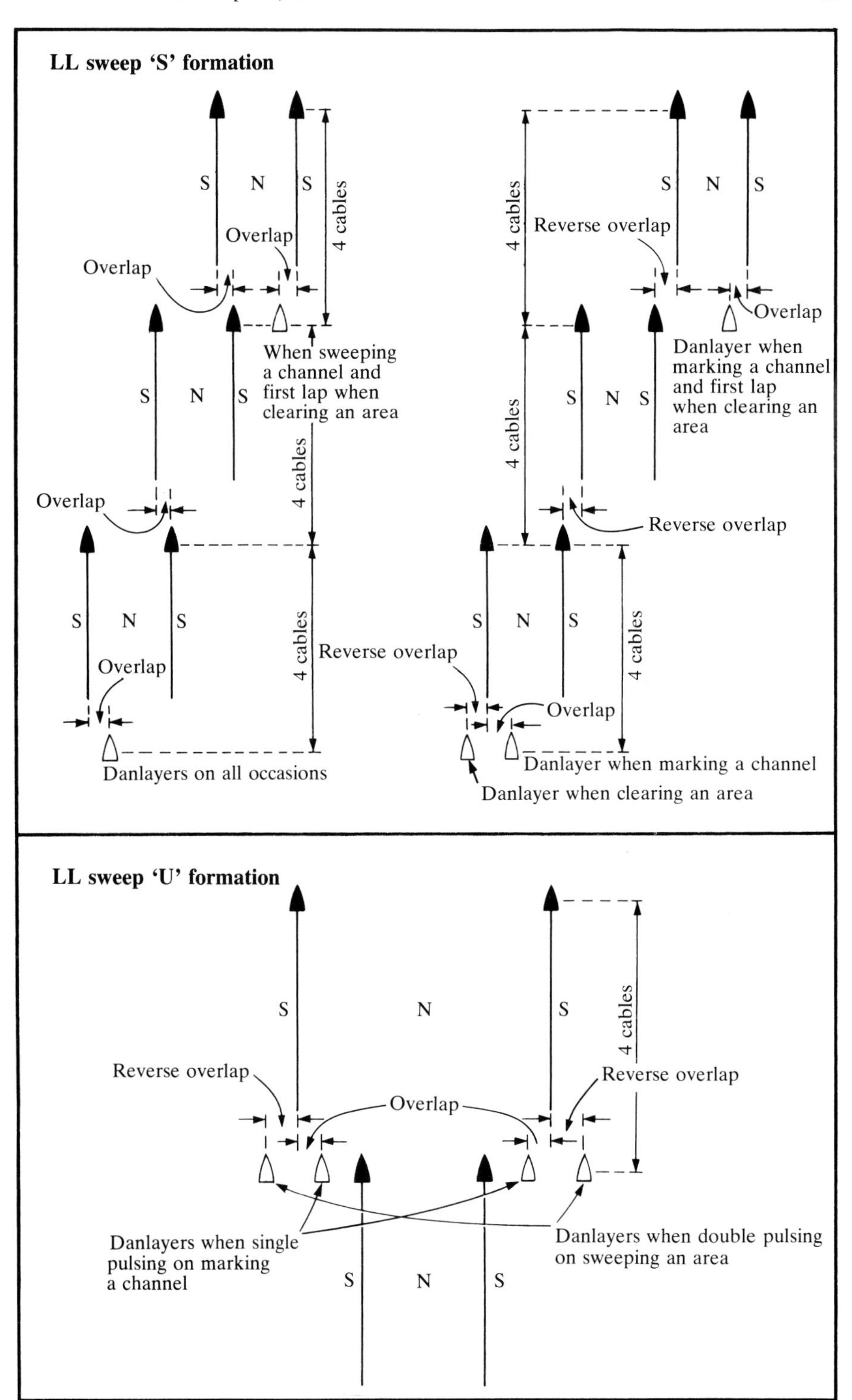
LL sweep 'S' formation
S N S
4 cables
Overlap
Overlap
When sweeping a channel and first lap when clearing an area
S N S
4 cables
Overlap
S N S
4 cables
Reverse overlap
Overlap
Danlayers on all occasions
S N S
4 cables
Reverse overlap
Overlap
Danlayer when marking a channel and first lap when clearing an area
S N S
4 cables
Reverse overlap
S N S
4 cables
Overlap
Danlayer when marking a channel
Danlayer when clearing an area
LL sweep 'U' formation
S N S
4 cables
Reverse overlap
Overlap
Reverse overlap
Danlayers when single pulsing on marking a channel
Danlayers when double pulsing on sweeping an area
S N S

Chapter 3

More sweepers—but more mines, 1941

The acoustic mine appears

Strictly speaking, it had already appeared in the last quarter of 1940, but the story can more conveniently be told here. German efforts to produce an acoustic mine went back to 1937. The first one was detected in the Thames Estuary in October 1940, when ships began receiving near-misses from mines exploding just ahead of them in areas which had been swept for magnetic mines.

Some top British scientists had said that such a mine was impractical but, early in 1940, HMS *Vernon* quietly started some research on counter-measures. It was as well that they did for, when the mines appeared, they had already built up a useful data bank on the problem.

It was not easy to produce an acoustic mine—one which was exploded by the noise of a ship's propellers reacting on a diaphragm in the mine. One of the first axioms of acoustic mining was that the greater the number of ship types against which it would be effective, the coarser the mine setting needed to be. However, this would mean that the proximity of the mine to the ship which it was striking would be less accurate, and it would therefore be easier to sweep.

So the first mines, set to catch slow-moving cargo ships, were fairly easily swept. But as the war went on, the acoustic mines were set more and more finely, and were designed to fire against particular types of ship. They were even being set not to explode under the sweepers, but to wait to catch the slower-moving cargo ships or large warships following astern of them.

Early sweeps

The mining school had done much good work in registering and analysing the 'sound signatures' left by the propellers of differing types of ships. Three frequency ranges were identified—subsonic, audio, and supersonic—and experiments started with the idea of producing effective sweeps against each of these. Unfortunately, the Clarence Pier at Portsmouth, where much of this work was being done, was destroyed in an air raid in March and many of these vital records were lost.

As soon as it became clear that acoustic mines were indeed being laid, experimental sweeps were quickly out at sea. Speed restriction and gunfire were the first emergency initiatives. It was hoped that, by steaming at slower speeds, ships would alter their 'sound signature' sufficiently to avoid setting off the mine. This never really worked (though it was effective against the pressure mines, nearly four years later). The use of gunfire was, surprisingly, quite effective.

Bursts of Lewis-gun fire were sprayed on the water ahead of the ship, and this rendered quite a few of the mines passive because they had been fitted with an anti-counter-mining device which picked this noise up. But machine-gun ammunition was in short supply in those days and needed for countering the dangerous air attacks, so this method was not encouraged.

Then came some amusing experiments in producing what was called 'white noise'—covering a wide audio range. Among the devices tried was an ordinary air raid siren, towed through the water in a steel torpedo-like casing, and a steam siren was also tried out in the same way. These were known officially as 'the cow' and 'the bull', and another variant as 'Basham'! Then pipe noisemakers were tried—pipes fitted in a loose framework which allowed them to chatter as they were pulled through the water. The framework became pretty complicated in some versions, and these were known as 'the palace of engineering' and 'the prefabricated house'. The most promising of them—it produced the loudest underwater noise—was called 'clanging Queenie'.

Other experiments included a flywheel, deliberately set to run out of balance, and a collection of tennis balls rumbling around in a steel box. But all of these 'good tries' were abandoned, since the noise they produced was in the subsonic range and therefore too broad to catch the mines.

The 'pneumatic hammer' was a promising line of research, work on which was already under way in mid-1940. When the German acoustic mine appeared this sweep was quickly at sea under test, and equally quickly achieving successes. With various modifications, this was the standard effective sweep against the acoustic mine. This hammer was an ordinary pneumatic road drill; the commercial name of the brand used was 'Kango', and so the device became known as the 'Kango sweep'. The manufacturers of the drill were surprised, but happy, to get large orders for their drills from the Royal Navy.

These hammers gave out a white noise without wide gaps in the sound range, and so set off most of the acoustic mines. The successful answer to the early acoustic mines was to have a sound source giving a loud and widely distributed noise on all frequencies, to fire the mine anywhere in the frequency band in which it was sensitive, and this the hammers did.

At first they were mounted inboard, with the hammer held up against a steel diaphragm, bolted over a hole in the bulkhead of the forepeak, and with the forepeak itself flooded. But two sweepers fitted in this way were sunk by acoustic mines in suspicious circumstances, so the hammers were then fitted inside a steel conical box. This box was fixed on a boom over the sweeper's bows, being lifted out of the water when not in use, and lowered down below the sweeper's forefoot when the sweep was being operated.

By February 1941, 60 sweepers had been fitted in this way and they were having enough successes for the acoustic mining campaign to be reasonably under control.

Later acoustic mines incorporated arming delays, and pulse delay mechanisms, as in the magnetic mines; and this simply meant that the sweepers using the LL (magnetic) and SA (acoustic) sweeps had to sweep the same patch of water many times, before it could be declared safe.

Not until 1943 was a really different type of acoustic device introduced, and by then the sweepers were ready for it and, with adjustments to their gear, they easily swept the new mines. The last technical advance in the acoustic sweep was to tow the hammer, in its same conical box, from a short boom amidships in the sweeper. This did away with the clumsy boom over the bows.

The 'Fessenden oscillator' was another early type of acoustic sweep. Similar in operation to the 'Kango' hammer, it consisted of a large diaphragm, up to 36 inches in diameter, fitted to the hull inside the ship, one on each bow, and producing a beating sound similar to that of the hammer. This was widely fitted in warships larger than sweepers, including battleships and cruisers, and continued in use until 1944. Ships with this sweep in operation were hit by acoustic mines, and it became clear that this oscillator could not keep up with the refinements in the settings of the mines. But in 1941 and 1942, many Nore Command destroyers were fitted with this sweep and achieved quite a few successes with it.

As with the magnetic mine just a year earlier, the break-through for the sweepers came with the recovery of an acoustic mine on shore, in November 1940. This disclosed details of the microphone fitted in the mine and the frequencies on which it operated. The hammer and the oscillator were thus confirmed as the two viable sweeps.

The explosive sweep came a little later, but can be covered here. Designed to catch acoustic mines which needed a build-up of sound to explode them, this sweep consisted very simply of a steel tube, down which hand grenades were dropped into the water, to explode at carefully calculated intervals. It was found that the steel tubes used ashore to mount flashing yellow lights on road pedestrian crossings were ideal for the purpose!

A simple vertical blast screen (against prematurely exploding grenades) was fitted on the after rails of the sweeper, and the tube was pushed through it. Through the tube were ejected the hand grenades, by a wooden ramrod. They were interspersed with wooden dummies to give the desired build-up of sound.

The grenades exploded some 40 feet astern of the sweeper, at a depth of about 20 feet. Quite frequently, they jammed in the tube, so giving rise to alarm and despondency all round, but otherwise it turned out to be quite an effective sweep.

On with the sweeping

The acoustic mining campaign began in the last three months of 1940, with acoustic mines being identified as the cause of explosions in East Coast and Thames Estuary areas. But the first really big acoustic lay, by German aircraft, turned out to be one of their biggest minelaying efforts of the entire war. On the night of December 12–13, a large number of minelaying aircraft flew over Southend, at the mouth of the Thames. They kept clear of the balloons flying above the convoy anchorage, and dropped many mines, most of them acoustic, inside the defensive boom. Forty-five were reported by mine watchers ashore alone, and this was a deliberate attempt to close the river.

The mines had been fitted with arming delays and, for four whole days, none exploded. Then ships began to be mined one after the other and many mines exploded spontaneously around them. Only three acoustic sweepers were available and one of them was damaged as she exploded her third mine. But by the evening of the fourth day the water inside the boom had been swept and a southbound convoy of 50 ships, which had been held outside, was brought in with everyone treading on tiptoe. Thirty-eight ships entered through the gate unharmed, then the next two were mined right in the entrance to the boom gate;

seven ships altogether were sunk that day in the same small area, and most became dangerous wrecks in the war channel.

A hectic winter

Very intensive efforts were made to bring into service more acoustic sweepers in addition to the LL magnetic vessels. The LL and SA sweeps were from this time on combined in the same ship, with the wire Oropesa ships not carrying influence sweeps (except in the large fleet sweepers). By mid-January, 224 SA sweeps of all types were at sea, and one sweeper put up 14 acoustic mines in four hours.

Moored minefields continued to be laid in the Dover Straits by German torpedo-boats. Two fields laid in January were swept for 17 days and 31 mines were discovered. The sweepers were continually attacked by aircraft, in spite of the bad weather, and many sweep obstructors were found to be mixed with the mines—58 sweeps were parted in this operation. A further 38 mines were washed ashore on the Kent coast from these fields, in the stormy seas.

Two German destroyers laid moored fields in March, between Dungeness and Selsey Bill on the South Coast. Three coasters in a convoy, and two minesweeping trawlers, were bagged here by the mines. The First Flotilla, of 'Halcyon' Class fleet sweepers, was called in to clear these fields, in thick fog and stormy weather.

The minesweeping fleet grew during these winter months, but it also had its setbacks. Twenty-three of the new MMSs, all fitted with both LL and SA, came into service by the end of March. But by that time German aircraft had sunk 26 sweepers, and the two flotillas of 'Halcyon' Class fleet sweepers had to be withdrawn from sweeping duties to reinforce the hard-pressed Western Approaches escort vessels. For a while, until the new 'Bangor' Class fleet sweepers came into service, four old 'Town' Class ships based at Harwich were the only fleet sweepers available in this priority mine warfare area.

Mines were laid up the Thames, near the docks, during this period. A big force

The stern of a 'Town' Class fleet sweeper, showing the 'A' sweep gallows from World War 1, with the Oropesa floats handled by light davits.

of shore minewatchers was recruited and tugs towing LL skids were sweeping the river without a break, three hours each side of high water. One of these tugs was sunk by a mine off the Ford dock at Dagenham.

The total of mines swept grew and grew, and the mix of mine types told its own story. By the end of April, the count stood at 1,044 magnetic, 849 contact moored, and 484 acoustic and, on April 11, the front-line Nore Command sweepers bagged their 1,000th mine.

Destroyer sweepers succeed!

One surprising and exciting feature of the early acoustic sweeping campaign was the extensive use of fleet destroyers off the East Coast, and in the Thames Estuary, as fast sweepers. They were fitted first with pneumatic hammers, inside each side of their stem, then later with the Fessenden oscillators, which were thought to be more effective in these ships. The destroyers then steamed fast up the swept channels, and had a surprising number of successes—records show the following, and no doubt there were others:

Tynedale, a 'Hunt' Class ship, got one in May 1941, with the hammer box on a hinged boom over her bow; *Holderness*, of the same class, put up four in July, while *Intrepid* also got four in that same month. *Windsor*, of the old 'V & W' Class, got one in the Thames soon after, while in August *Holderness* got two more. But her sister ship *Quorn* was damaged off Harwich by a mine which went off 50 yards away, while her SA gear was out of action. In October, the fast-sweeping *Holderness* put up another one and, not to be outdone, two V & Ws, *Versatile* and *Wallace*, each exploded one before the year's end. By September, 11 Nore Command destroyers had been fitted with SA sweeps and, by December, two of these destroyers were sweeping ahead of all the East Coast convoys.

Summer progress

The pace was maintained right through the summer months, but with the sweepers keeping on top of the mines.

The new MMSs continued to come into service in numbers, and the first of the Canadian-built units of the class arrived via Iceland in September. 'Bangor' Class fleet sweepers were also being rushed into service, and this augured well for the clearance of known fields, though their speciality was to be wire sweeping.

Sweeper casualties mounted in parallel. MMS 39 was sunk off the Thames when her SA sweep broke down, and MMS 40 was badly damaged when she put up an acoustic mine only 40 yards off her port bow.

Nevertheless, notable sweeping successes were turned in as well. By June, the trawler *Arctic Hunter* had scored her half-century of influence mines swept in the Thames Estuary and, in the same area, four LL/SA drifters bagged 11 acoustic and eight magnetic mines during one hour in August. Not to be outdone, four MMSs six weeks later cleared 11 acoustic and one magnetic in as short a period.

Already there were changes in sweeper policy. The MDSs were taken out of service in mid-year, to be converted to LL sweepers, and SA sweeps were being fitted to a fleet carrier, a battleship, and several cruisers.

The year closed with a big minefield laid in the Dover area in early December, and it took most of that month to clear it. The Dover MTBs failed to intercept some German surface minelayers, and the clearance took 29 sweeping days, during which four trawlers and a drifter were lost or damaged. As many as 60 sweeps were parted, and fighter protection was needed throughout. The weather did not help—four sweeping days were lost to gales and another eight due to strong tides, which made accurate sweeping impossible.

The Mediterranean sweepers score

The German mine warfare campaign in the Mediterranean was gaining pace, and the sweepers based at Malta had an especially busy time. In February, German aircraft mined the harbour on 11 successive nights, and the whaler flotilla and the skid-towing ships there were kept hard at it. Four minesweeping 'Flower' Class corvettes were passed through to Malta with one of the famous convoys, and they stayed on performing magnificent work until, with the other warships based there, they had to move east while the German aircraft retained control of the skies.

In April *Gloxinia*, one of these corvettes, fitted for LL and SA, lifted a magnetic and an acoustic off Tobruk, and later that month, she detonated 14 ground mines inside Grand Harbour itself. Then she was damaged by a near miss. But the new fleet destroyer *Jersey* was mined and sunk in the harbour entrance on May 2, though the sweepers worked flat out. Every suitable craft was pressed into service, including small tugs and a hopper barge, which lifted seven magnetic mines before another one sank her.

Other sweepers ranged far and wide around the Mediterranean. A 'Hunt' Class fast sweeper was sunk by German aircraft off Mersa Matruh, on the North African coast, at the end of January, while during the fighting for Greece, *Salvia* and *Hyacinth* of the 'Flower' Class, bagged five magnetics off Piraeus, even while the Greek Navy was being immobilised to avoid capture. *Salvia* was then damaged in an air raid on the island of Crete, and four of the 'Flower' Class sweepers passed through the Suez Canal, for service in the Indian Ocean. But before the damaged *Salvia* could leave with them, she was torpedoed off Tobruk.

Two years' sweeping

A quick look at the results, two years after the European war started, tells its story. In UK waters, two of the 'Halcyon' Class fleet sweepers were lost, and two of the 'Hunt' Class; but 19 of the new 'Bangor' Class entered service in 1941. Nine of the paddle sweepers, and three mine destructor ships, had also been lost.

The trawlers—now here was a triumphant picture! From 40 Oropesa trawlers available at the outbreak of war, the sweeper fleet grew in two years to 262 Oropesa trawlers, 166 LL/SA trawlers, 45 motor minesweepers, and over 100 skid-towing drifters. The losses in this growing fleet tell their own tale, too—74 trawlers, one MMS and six drifters had been lost.

Abroad, four 'Town' Class sweepers were lost, but 15 of the new 'Bangor' and Royal Australian Navy 'Bathurst' Classes joined the sweeping force. Trawlers grew from virtually none, to 177 Oropesa-fitted ships, plus 58 LL sweepers.

Casualties other than in sweepers showed how mine warfare quickly became a critical factor. At home, eight warships were sunk, and 18 more damaged; and 274 merchant ships were sunk by mines, and 104 damaged. Abroad, losses were much fewer, away from the hard-fought East Coast of England. Three warships were sunk and one damaged, plus 17 merchant ships sunk, and seven damaged. But the mines swept made up for all that—the bag showed that the swept channels had been kept open—sometimes by a hair's breadth, by the devotion of the sweepers, and the unceasing improvisations of the technologists.

	Magnetic	Acoustic	Contact
UK waters	1,262	962	818
French waters	47		2
Mediterranean	63	28	71
Suez Canal	28	2	
Australia			36
New Zealand			89
Totals	1,407	992	1,042

German acoustic mines

Here are some technical details of the German mines described in this chapter:
A.1; this was the first acoustic mine, introduced in October 1940. It had a coarse setting and so was fairly easy to sweep. It also had a six-day arming delay clock. The sweep SA Type A Mark II was especially effective against it.
A.2; this one appeared in July 1941. It incorporated increased sensitivity, a pulse delay mechanism with up to 12 actuations needed, but no arming delay was included. It reacted to audio-frequency sound of medium intensity, but suffered from the disadvantage of still not firing close enough to the target ship. The SA Type A Mark II swept it well.
M.A.1; (British name 'Sammy'). This was a new mine and altered the whole British sweeping policy when it appeared in September 1941. It incorporated for the first time a combined magnetic-acoustic unit, and it was necessary to have sufficient acoustic energy present during the operation of the magnetic unit for the mine to fire. The microphones had wider listening ranges, and 'fixed' more accurately on propellers and engine rooms. The sweepers needed to sweep a much larger area of seabed, to be sure of lifting this mine, but the combination of LL and SA Type A Mark II usually swept it.
M.A.101; this was the next new mine, another combined magnetic-acoustic, and it appeared in the Mediterranean in September 1943. It had four non-resonant microphones, in parallel, and they needed a build-up of sound to close the acoustic contacts. The acoustic side needed to be active for 15 seconds to trigger the magnetic side and, if the mine did not then fire, the acoustic side became passive again. The spring hammer SA sweep got it, when combined with the LL sweep.
A.104; appeared at the same time, but in UK waters. It was a new type of plain acoustic mine, with a pulse delay mechanism needing up to ten actuations. This device was of improved construction and it was designed to discriminate between the noises of ships and of acoustic sweeps. It would fire only on a rapid increase in sound, such as exists under a ship. It was an unsuccessful mine, but it gave warning of the trend in German mine design, and the sweepers had an antidote ready when an effective mine of this type did appear.
A.105; appearing in April 1944 was that mine. It was the final development of the three-relay microphone system, and continuous sound sources, such as a hammer, simply rendered the mine passive. Pulsed sound was necessary, and the explosive sweep, in its two marks, was the only one to lift this mine.

British acoustic sweeps

The following are the official designations of the various sweeps touched on in the narrative:
SA Type A Mark I—Kango hammer, fitted internally. It was held up to a steel

An acoustic sweep, the Kango hammer Type SA, a Mark IV, being recovered from an MMS of the Royal Navy. The davits were mounted on the foredeck.

diaphragm of 19 inches diameter, in the flooded forepeak of the sweeper. This was the earliest sweep, in early 1941, and was discontinued when the external box came into service.

SA Type A Mark II—Kango hammer, the first external fitting. It was housed in a conical steel box, which was fitted on a hinged boom, over the sweeper's bow. It operated 12 feet below the surface. Some of this type were in operation up to 1945.

SA Type A Mark III—Kango hammer, fitted internally. Two hammers were used in this version, each held up against a steel diaphragm 24 inches in diameter, one on each side of the stem, in a flooded forepeak. Used in 1941 and 1942, but discontinued by 1943.

SA Type A Mark IV—Spring hammer, in a towed box. It was towed from slings below the sweeper, and handled from a short boom amidships, or in MMS, from the boat's davit. It became the standard SA sweep in new fleet sweepers, BYMSs, and MMSs, from the end of 1942 onwards.

SA Type A Mark V—Spring hammer, towed as in the previous mark, but using 'rabbits ears', to divert the box away from the ship.

SA Type B—Probably an early experimental type, but research has not thrown up any details, and it does not appear to have been used operationally.

SA Type C Mark I—Fessenden oscillator, fitted internally. It was fitted inside the stem, in a flooded forepeak. The diaphragm was of 36 inches diameter. Used in destroyers and some sweepers, up to 1942.

SA Type C Mark II—Two Fessenden oscillators, fitted internally. One was fitted on each side of the stem, in a flooded forepeak. The size was as in the previous mark. Fitted in some large warships, and in a number of destroyers, up to 1944.

SA Type D Mark I—Kango hammer, towed overside in a steel box, and producing a special range of 'white noise'; an experimental type, this one was soon discontinued.

SA Type D Mark II—Pipe noisemakers, as described earlier. Discontinued at an early stage.

SA Type D Mark III—Another type of experimental pipe noisemaker, and also discontinued before entering operational service.

SA Type E—Hand grenades, thrown overboard indiscriminately—not as in the explosive sweep, below. Discontinued very early.

SA Type F—The use of machine-gun bursts in the water, to render the mines passive. Surprisingly successful; discouraged early in the war, due to shortage of ammunition, this method was used again in mine clearance when very sensitive acoustic mines were mixed in with other types.

SA Type G—Spring hammer, in a steel box, as before, but this time towed from the LL tails, 400 yards astern of the sweeper. A later refinement in SA sweeping, and used especially in postwar mine clearance.

Explosive sweep Mark I—A sweep using standard hand grenades, fired by being pushed through a steel tube, mounted on the rail near the sweeper's stern. It came into service, as did the later marks, in 1944, but is described here for easy reference. The firing tube was ten feet long, and of 2½ inches diameter. The firing sequence was: one grenade, one dummy, two grenades, one dummy, ten grenades; the wooden dummies were included to produce the required build-up of sound. The firing sequence took two minutes, with five-second intervals between grenades. The grenades had a four-second firing delay from the release of the pin, and the explosions occurred (one hoped) 40 feet astern of the sweeper, at a depth of 20 feet, when the sweeper was steaming at 7–8 knots. Each grenade contained a charge of three ounces of Barratol.

Explosive sweep Mark II—As Mark I, but for use against coarse-set mines. The firing tube was 24 feet long, and was held vertically in the sea to a depth of three feet. Twenty detonators (not grenades) were stored in the tube before firing. The firing process was to disengage a plunger at the bottom of the tube by means of a lanyard, and the detonators were fused electrically when they passed a contact at the end of the tube on their way out. The detonator charge was 0.58 grams. A slightly less hazardous sweep than the previous one, but there was not much in it!

Explosive sweep Mark III—Again, similar to the Mark I, but with specially-made charges, carefully graded; the firing sequence was two charges, one blank, three charges, one blank, nine charges. Each charge contained four ounces of explosive, and was in a small metal cylinder, three inches long, and two inches in diameter. They were contained in a clip in the discharge chute, and the ramrod was by this time mechanically driven. All charges had to be discharged in a period of 3½ seconds, precisely, if the sweep was to be effective.

Motor minesweepers (RN and RCN)

Prewar design

In 1937, the Admiralty had foreseen the need for a specialised design of inshore sweeper, and two experimental vessels were commissioned from John I. Thornycroft Ltd, from their Hampton-on-Thames yard. These were small sweepers, based on the coastal fishing boats of the day. Of only 32 tons displacement (the later MMSs were to be of 165 and 255 tons, respectively), they had a total length of 75 feet, only half that of the later long MMSs. They had three propeller shafts, with engines giving 15 knots running free, and 10 knots while sweeping—much higher than subsequently achieved. They were of roughly the same dimensions as the wartime HDMLs, which were later seen to be too small for coastal work.

The minesweeping gear in these two ships seemed almost to take a secondary position, perched right on the stern. But they were said at the time to be efficient sweepers; they were sold to Turkey in 1939. One other experimental MMS was built before the war, MMS 51, and this followed the lines of MTBs currently under construction. But she proved also not to be the answer to the inshore minesweeping requirement and, on the outbreak of war, this boat reverted to an MTB configuration.

The wartime requirement

Then, in 1940, when an effective sweep was quickly developed against the magnetic mine, the need was seen for a new and specialised sweeper. This would be built of wood to minimise the magnetic danger, and would be specially designed to handle the new LL sweep, with its big reel aft, and the need for a clear run for the cable to avoid chafing over the stern.

In this quick design many of the old fishing characteristics of the trawlers were retained; it is important to remember this when comparing the MMS design with that of the YMS, which was started from scratch. One should also remember that

Left *A 'short' MMS steaming through a swell. The SA hammer and boom stand out, and the clear sweep deck aft for the LL cables.*

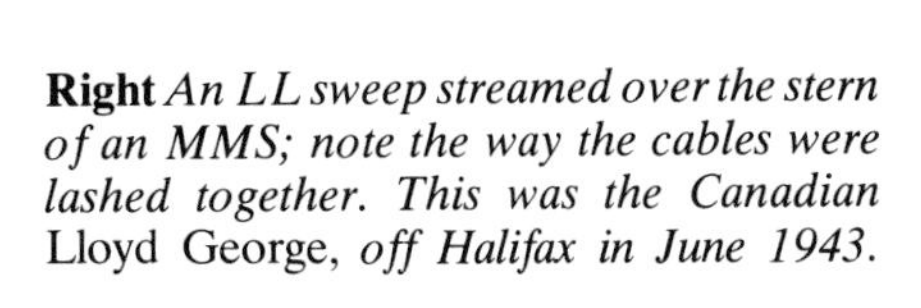

Right *An LL sweep streamed over the stern of an MMS; note the way the cables were lashed together. This was the Canadian* Lloyd George, *off Halifax in June 1943.*

the MMS was built in Britain at the height of the wartime shortages, so that only a single shaft was included, giving a sweeping speed which was too low, and the armament in the initial 105-foot version was really minimal.

The first version of the MMS started to enter service in 1941. It was much safer than the steel trawlers for sweeping magnetic mines, but its towing capacity was clearly poor—its maximum load was 1½–2 tons, at 8 knots—this was adequate for the LL sweep, but not so good for wire sweeping. The acoustic mine appeared while these early MMSs were pouring out of the builders' yards, and the SA hammer box was quickly added, mounted on the boom which lowered over the bow. In spite of their limitations, this class largely took over the LL and SA sweeping around the British Isles and did a splendid job. This is demonstrated by their war losses—15 ships in 1942, and when the influence mining campaign hotted up again in 1944, another eight.

A need was soon seen for a larger, more seaworthy version of the MMS, and this appeared in the 126-foot version. These were intended for sweeping the war channels in any weather, with a stronger anti-aircraft armament. They had better living accommodation for the hard-pressed crew, and the same sweeps as their predecessors, the LL Mark III and the SA. But their main drawback was that they received the same single shaft and engine as their shorter forebears, which meant in service that both their towing ability and their manoeuvreability was worse than in the 105-foot version. This was an indication of wartime engine shortages, since the design intention was that they should be twin-screw ships, with a performance comparable with the American YMS, coming out in parallel. It is interesting that in these sweepers, just as much as in a comparison of the frigate and destroyer escort classes of the period, we see the shortage of suitable engines limiting the effectiveness of these important vessels.

The longer class came out in 1942 and missed the worst of the mining campaign; this is reflected in the war losses of the 126-foot class—only two ships.

Canada made a significant contribution to the building programme of both these MMS classes. Ten of the shorter boats were built in Canada and, of these, six went to the Mediterranean, and four to the East Indies Fleet. They had hair-raising delivery voyages; they were given a minimum endurance of 2,000 miles, by the installation of deck tanks which took an additional 36 tons of gas

Above left *One of the Motor Minesweepers built in Canada for the Royal Navy. MMS 200 is seen iced in at Shelburne, Nova Scotia, in February 1943, before sailing for Europe.*

Above *A long MMS (127 feet), No 1044. This type was an interesting comparison with the American YMS, of about the same size.*

Below *The LL cables being returned to their reel on an MMS. This was the Canadian* Lloyd George *off Halifax in June 1943.*

Above *A kite/otter being hoisted out in the Canadian 'Bangor' Class fleet sweeper* Stratford, *off St Johns in November 1943. A cutter is clamped on the sweep wire, just below the otter.*

Below *The Canadian 'Bangor'* Malpeque *streaming a double Oropesa sweep in the English Channel in July 1944. Both floats are away, and the kites are following.*

oil. Also a 'jury rig' was fitted, a suit of sails giving 6 knots in a fair wind! Two of the 'Bangor' Class fleet sweepers completing in Canada were earmarked to escort them across the Atlantic to East Africa, the original route via Ponta Delgada in the Azores being rejected as too boisterous, and they eventually voyaged from Trinidad to Lagos.

Of the longer, 126-foot version, Canada built more units for the Royal Navy than she built for herself. Then, of the 16 units built for the Royal Canadian Navy, ten were cancelled as the minesweeping requirement by that time had diminished, and the other six were completed and transferred on commissioning to Russia.

The list of ships of this class transferred to Allied navies is worth some study; in early 1945 when Parkeston Quay, the naval port at Harwich, was crowded with large numbers of MMSs, it was noteworthy that the ensigns flying beside the quay were as varied as the accents of the crews.

	Short MMSs	**Long MMSs**	**BYMSs**	**'Isles' Class trawlers**
Built	1940–44	1943–45	1942–44	1941–45
Displacement in tons	165	255	207–215	545
Length, total in feet	119	140	136	164
Length, between posts	150	126	130	150
Breadth in feet	23	26	24½	27½
Draught in feet	9½	10½	6	10½
Engines	Diesel bhp 500	Diesel bhp 500	Diesel bhp 1000	Reciprocating ihp 850
Shafts	1	1	2	1
Speed, free, in knots	11	10	14	12
Complement	20	21	30	40
Armament			1 × 3″/50	1 × 12 pr.
	2 × 0.5″	2 × 20 mm	2 × 20 mm	3 × 20 mm
No built	295	96	142	135
No cancelled	23	10	0	0
War losses	34	2	8	13
Built in Canada	26	14		
Built elsewhere				
Nassau	4			
Grand Cayman	2			
Colombo	3			
Tel Aviv	2			
Beirut	2			
Lost on the stocks				
Singapore	9			
Rangoon	10			
Hong Kong	4			
Cochin	9			

Transferred to Allied navies

Short boats: 9 to Belgium, 7 to France, 5 to Holland, 3 to Russia, 1 to Greece.
Long boats: 9 to Holland, 6 to Russia, 2 to Norway.

'Bangor' Class fleet minesweepers (RN, RCN, and RIN)

This was the first new class of fleet sweepers built for the Royal and Commonwealth Navies during the war. It was considerably smaller than either its predecessor, the prewar 'Halcyon' Class (245 feet total) or the later 'Algerine' Class (225 feet), but this class, at 162 or 180 feet (there were several versions) represented good value at the time. They were very useful Oropesa sweepers and did excellent service in this role, especially in the waters around the British Isles; but when influence sweeping by fleet sweepers became a necessity, their cramped deck lay-out emphasised the need for a larger ship. Fortunately, this need had been foreseen by the Admiralty long before it became essential and, by that time, the larger 'Algerine' Class were appearing.

The design followed naval lines, and these ships have a distinctly professional appearance. The hull seems a diminutive of the 'Halcyon' Class hull and the bridge, while much smaller, has with the funnel that same 'naval' look. It is noteworthy that this was just about the first covered bridge in the Royal Navy, in a ship of this size. Soon after the war it became normal but in 1940 and 1941 some old hands complained that the bridge roof prevented them seeing the attacking aircraft clearly enough!

There was a pole foremast, a short mainmast and a 27-foot whaler was carried on the port side, which looked disproportionate in size to the ship. A single 12-pounder A/A gun was carried on the foredeck, and either a quadruple 0.5-inch (not a very effective gun) or a single two-pounder pom pom aft, just above the sweep deck. Two single 20 mm Oerlikons were added at the sides of the bridge, when the supply of these weapons became a little easier.

The sweep deck was crowded, even for wire sweeping; one result was that this class was only able to carry a few depth charges in single traps round the stern, plus two, rather than the normal four, depth-charge throwers. They were not normally used on escort duty by the Royal Navy, but they were extensively used as such in the North Atlantic and in Canadian coastal waters by the Royal Canadian Navy. In that role they performed very well, especially in coastal waters. In a number of these units the minesweeping gear was removed while on escort duty, partly to give greater space on the after deck and partly to avoid serious maintenance problems with the sweeping gear when the ships were operating in the arduous North Atlantic weather. This class did not have very good radar equipment. Type 290 was fitted in most at the masthead, but they never received the heavier Type 271.

Production

For the Royal Navy a total of 60 ships of the class were built—54 in the United Kingdom, and six in Canada. Of these nine were transferred, after some RN service, to the Royal Indian Navy. It is noteworthy that no ships of this class were cancelled in the UK, probably due to planned progression to 'Algerines' Class in 1940. Six ships were lost due to enemy action.

In Canada, 54 ships were built for the Royal Canadian Navy. Of these, five had been ordered by the Royal Navy but were loaned to the RCN on completion. It is interesting that the Royal Navy allocated Canadian names to these ships, as well as to those which were commissioned by the RN itself. Five of the RCN units of the class were lost by enemy action.

A further four units were built in India, for the Royal Indian Navy. The latter thus had a good force of these ships. Four more units which were under construction in Hong Kong and Taikoo, were lost during the Japanese invasion.

Chapter 4

The battle goes on, but America joins in, 1942

United Kingdom waters

A hard, slogging year followed for the sweepers at home; a bad winter combined with unrelenting enemy minelaying kept them under pressure. By now, though, they were getting the measure of the campaign. There were no more real surprises in the magnetic mine (the sweeps had become quite sophisticated), the acoustic mine had been successfully countered in the previous year and, in this winter, the combination magnetic-acoustic mine also appeared, but effective counter-measures were soon taken. The problem was to make the best use of the growing number of sweepers—and especially of the MMSs—while modifying the basic LL and SA sweeps to meet the differing sensitivities of the mines laid.

New ships coming into commission greatly strengthened the arm of the sweeping forces. The first of the new 'Algerine' Class fleet sweepers was commissioned in March, and the Twelfth Flotilla was operating by the end of summer. The first three of the longer, 126-foot MMSs came into service in December, and there were 100 of the two types by the year's end. All these new ships were fitted with the latest types of sweep.

Ships built in North America were arriving in strength, but not without difficulties on the stormy Atlantic crossing. MMS 13 broke down in mid-ocean on her maiden voyage from Canada in October, and needed a tow to the United Kingdom. The same happened to two out of four of the new BYMSs, coming over in company; the big rescue tugs soon pulled them to safety.

Two long MMSs of the Royal Navy, MMS 1022 and another, sweeping for ground mines in a rough sea off the East Coast of England.

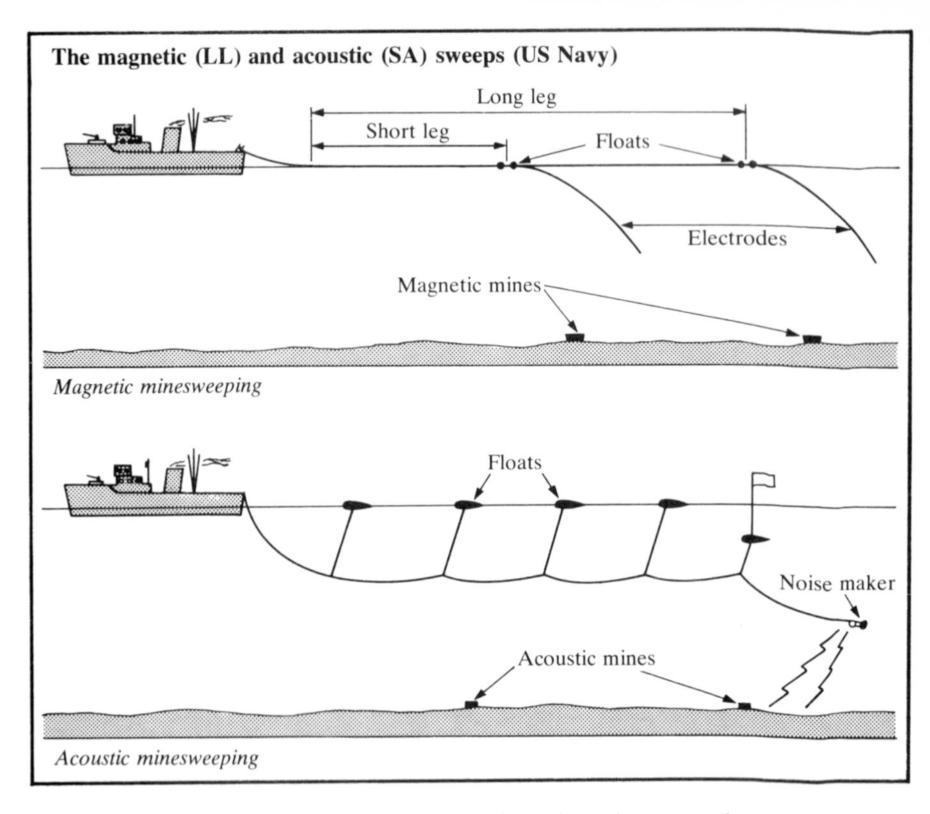

The magnetic (LL) and acoustic (SA) sweeps (US Navy)

The auxiliary paddle sweepers were being phased out, as the new 'Bangor' Class grew in numbers; the latter totalled 33 in this year, and the Seventh and Eighth Flotillas of paddlers were converted to auxiliary anti-aircraft ships.

There were also 20 'Flower' Class corvettes, and six 'River' Class frigates in commission, and fully fitted as sweepers. The new escorts were badly needed in the North Atlantic; those based at Londonderry were designated as stand-by minesweeping flotillas and, as far as the records show it would appear that apart from trials and working up, they were never used on that duty. This has a parallel in Canada where the 'Bangor' Class sweepers commissioning were used almost entirely as Atlantic escorts, and the minesweeping equipment in quite a number of them was actually landed to make more room for the depth-charge equipment.

From this point on the totals of operational sweepers did not grow significantly for, as the new MMSs and 'Isles' Class sweepers came into service, quite a number of the older and slower trawlers and drifters were paid off. They returned to their commercial owners—many had been built before World War 1.

E-boat sorties to the fore

In this year, with the British Coastal Forces and destroyers now able to defend the coastal convoys more effectively, the E-boats turned their efforts to laying mines. Their favourite tactic was to lay on a fine winter's night, which was likely to be followed by several rough days, on which the LL trawlers and the short MMSs would not be able to operate. They were using magnetic-acoustic mines with increased sensitivity on the magnetic side which trebled the danger area for the British ships.

The sweepers, however, were undaunted—the action report of one Humber trawler group recorded that, 'Oropesa patrols have been carried out continuously, but unfortunately, no contacts were made with E-boats!'

The Germans made some mistakes in the use of their more sensitive mines. First, they allowed the E-boats to use their wireless and radio-telephone freely while laying, so that their positions could be carefully plotted ashore on the East Coast and thus the location of the new minefields identified. Then they laid the mines too close to the swept channels, and used varying clock delays. The result was that the daily routine sweep of these channels often swept some of the earlier ripening mines and so revealed the presence of the others. In April, E-boats laid their first fields of moored mines for 18 months—four small fields were found containing 80 mines.

More than 1,000 individual minelaying sorties were flown by German aircraft on moonless nights during this year. They were protected to some extent from shore anti-aircraft fire because their new, more sensitive mines were effective in deep water and they did not need to fly over the inner anchorages in order to lay them.

The sweepers were still under air attack by day, however, and on moonlit nights. Thirty-seven separate attacks were made on trawlers on the East Coast and five were sunk, though they brought down two of their attackers.

Mine bags and casualties

The measure of the sweepers' success lay not only in the safe arrival of the convoys, but also in the growing bag of swept mines. In 1942 only just over half as many ground mines were swept as in 1941, but the total of moored mines cut went up again, due to the E-boat lays. On average one ship was lost for every 100 mines laid.

There were still casualties to the sweepers. MMS 180 was lost by collision in February, showing that mines were not the only hazard. The destroyer *Quorn* put up an acoustic while travelling at full speed, but then a second mine got her and she had to be towed back to Harwich. Her sister ship *Cotswold* was also damaged by a mine soon after. The old 'Smokey Joe' *Fitzroy,* Senior Officer's ship of the Fourth Flotilla, was sunk by a mine in the East Coast mine barrage in June and several trawlers were lost to mines during the year. Breakdown of the SA acoustic sweeps at sea was often the cause. Diaphragms cracked, hammers failed and the sweeper crews had to change parts of the gear at top speed in heavy weather conditions.

However, the sweeping went on and on without a break. By mid-year, the Nore Command had swept its 2,000th mine of the war (by now, three-quarters of them were ground mines). An LL/SA trawler, *Delphinus,* had lifted 50 ground mines in seven months, mostly at night in the main Nore approach channels to the Thames. Much of the sweeping by now was done at night, ahead of the convoys, for the constant nightmare of the Nore Command was that the swept channels might get blocked by the wrecks of ships sunk by mines.

The headquarters ships based at the main minesweeping ports were now playing an important part. *Ambitious* was at Scapa Flow, and *St Tudno* at Sheerness; the latter, to disprove the sweepers' claim that she was aground on her own stack of empty tin cans, put to sea on a spring cruise in May and reported frankly that her ship's company had not stood up too well to the ship's motions in a seaway!

Minesweepers to North Russia

Early in the year, some fleet sweepers were needed to keep clear the approaches to Archangel, the arrival port in North Russia for the hard-pressed convoys. In January, *Bramble, Seagull,* and *Speedy* arrived from Scapa with one of the convoys, to be followed by *Leda* and, over the next two years, a flotilla of 'Halcyons' was stationed in the area, operating under very harsh conditions in the winter.

They worked hard, both as sweepers and as local escorts, averaging 20 out of every 28 days at sea. They swept with both Oropesa and LL sweeps, and achieved good results. But conditions were difficult for the crews—severe cold and ice on deck, combined with savage air attacks. The bulkheads in the messdecks sweated and dripped on them, and there were few recreational facilities ashore—though one rating was heard to observe, on his return on board, that he had never before experienced getting all he wanted for nothing, with a glass of vodka thrown in!

The sweepers acted as escorts to the ocean convoys, when on passage to or from Russia, and took a good part in those battles. In one skirmish, one dark winter's night in the Arctic, *Bramble* was sunk while defending her convoy against a German battleship and several destroyers. The name of this heroic ship was carried on to a new unit of 'Algerine', commissioned early in 1945, which led the Third Flotilla in mine clearance operations all round the British Isles, and up the Dutch coast, over the next two years.

Mediterranean mines

The campaign went on throughout this year, with the island of Malta as the main focus for the German and Italian forces. Continuous sweeping was carried out by the forces based there, which by now included four fleet sweepers, two 'Halcyons' and two 'Bangors', the latter, on their arrival in mid-year enabled the sweepers to keep the approach channels and the main harbour open.

In a period of six weeks in mid-year, 186 mines were cleared, 65 of these by the small drifters and harbour craft which had been pressed into service. There were casualties, too—the big minelaying submarine *Olympus,* leaving after a cargo-carrying trip, and crowded with passengers, was mined at night after leaving Grand Harbour, with very heavy casualties. *Hebe,* one of the island's fleet sweepers, was damaged, as were three destroyers and a merchant ship, when the convoy 'Harpoon' arrived in June. The 'Hunt' class destroyer *Southwold* hit a mine and sank just outside Marsaxlokk Harbour when assisting the damaged fast cargo ship *Breconshire* into port, after bombing.

Other sweepers continued their good work in the eastern part of the sea, especially off the embattled North African coast. *Snapdragon,* an LL-fitted 'Flower' Class corvette, was bombed in Benghazi harbour, and *Erica,* of the same minesweeping group of corvettes, was mined just outside the same port.

There were even two technical breakthroughs by the Malta sweepers in this year of the island's great blockade; and perhaps the most significant was that achieved by the smaller ships. The Third Motor Launch Flotilla, of 112-foot 'B' Class boats, had come through to Malta with the convoy in June. In helping to clear Malta's mine-infested channels and harbours they improvised an Oropesa sweep using a light sweep rope which had to be worked by hand. Shades of peacetime days, this had originally been devised for use by brass-funnelled picket boats to use round battleships in harbour, but now the idea proved its worth. They were successful with it, too—by November, they had cleared over 100 moored mines in the approaches to Valletta. Their successful initiative was picked up for assault landings from 1943 onwards, and indeed for all clearance operations by fleet sweepers, for two MLs carried out a skim sweep using this rope method, ahead of the leading fleet sweeper in uncleared waters.

The other notable event was a successful 'A' sweep, carried out by the four fleet sweepers based at Valletta—*Hebe, Speedy, Hythe* and *Rye*. They successfully swept a field of 24 moored mines in this way, and this was one of the rare occasions on which the 'A' sweep was used after the first year of this war. They fixed their position accurately by flags flown ashore and, as their turning circles were different and they were of different classes, this operation was considered to be highly creditable. The trawler *Beryl*, one of the Malta stalwarts, was used as a danlayer, and another trawler, *Swona,* for sinking the mines that had been cut.

South Africa made a notable contribution to minesweeping in the Mediterranean. Four LL magnetic sweepers were sent up from Durban early in 1942 and, after some refitting on arrival, were in service by May. They performed magnificently, especially along the battle-torn North African coast, and they were prominent in the operations at Tobruk when it was overrun by the Germans.

When the tide turned and Allied forces started to advance westwards again, the South African sweepers were again in the van. This little sweeping force remained in the Mediterranean, hard at work until the end of the war, and the last ships did not reach home until the end of 1945.

South Africa

Another field appeared off South Africa, in March 1942; this one was laid by the German *Doggerbank*, and was discovered when a Dutch ship was sunk not far from the end of the Cape Town searched channel. The mine clearance flotilla was quickly sent round from Durban and a total of 12 sweepers started a search of the area.

The water was very deep in the area, with strong currents, so that not only did the dan-buoys dip below the surface, making accurate sweeping difficult, but it was also suspected that some of the mines, swayed by the currents, were dipping below the level of the sweep wires. The flotilla swept convoys arriving at Cape Town in and performed the same service when they left and, although only a small number of mines had been laid, and in scattered groups, the minesweepers took great care in clearing the field in this congested area for shipping.

Far East

The Japanese advance was now causing the sweepers in this theatre some serious problems. *Hollyhock*, a 'Flower' Class corvette of the Third Minesweeping Flotilla, was bombed and sunk off Ceylon and, more strangely still, MMS 51, which had gone east soon after her completion, left Singapore on the evacuation, with the Australian 'Bathurst' Class minesweeper *Maryborough,* but later was caught at sea off Batavia by three Japanese cruisers with attendant destroyers, and sunk by close-range gunfire.

However, in this theatre of war, too, signs were already appearing of the changing tide. The Fourteenth Flotilla of 'Bangor' Class ships arrived at

A US Navy tug towing a magnetic skid sweep at Pearl Harbor, in August 1942. Salvage on the sunken battleship Arizona *is proceeding in mid-distance.*

Mombasa and, in Australia, the fine new sweepers of the 'Bathurst' Class, some built there for Royal Navy account but all manned by the Royal Australian Navy, were commissioning in numbers and building up quickly to their full strength of 60.

America joins the Great Alliance

Pearl Harbor brought the United States into the global conflict, but the Americans had started the revival of their minesweeping force some three years earlier. Minesweeping in the US Navy before 1938 was rather more in the doldrums than it had been in its counterpart, the Royal Navy, before 1939. The only sweepers available to them were the old 'Bird' Class left over from World War 1; and they had been designed like large tugs and so were hardly good enough for minesweeping service in World War 2. Some of them were still in service in the Philippines in 1942, but they did not play any significant part. Similarly, up to 1938 there was little interest in developing sweeps and ships, but the clearly growing clouds of war in that year sparked off a new effort at naval headquarters in Washington.

First, a new design of fleet sweeper was ordered in 1938, when the two initial ships of the 'Raven' Class were put under construction. They had to be redesigned in 1941, before they had been fully evaluated, once the magnetic mines appeared in the European war. But it is a remarkable tribute to that excellent design that it took its place with the British 'Algerine' Class as one of the two finest fleet sweeper designs of World War 2; yet the hull design and deck layout was finished before the war.

Then an emergency programme was put in hand in 1939 and 1940, to acquire as many commercial trawlers as possible for conversion to coastal minesweepers; this plan paralleled that in the Royal Navy in 1939 when hundreds of commercial trawlers and drifters were requisitioned. For the US Navy, about 60 ships, a mixture of deep-sea trawlers and purse-seine net drifters, were purchased and converted to sweepers. At first they were fitted only with wire sweeps, but when the LL sweep was introduced, following the Royal Navy's experience, a number of these ships were fitted for this type of influence sweeping instead. It is interesting that these vessels in many cases were fitted with a steel reel, stowed horizontally on top of their after deck, in place of the normal vertically-fitted reel. The reason was that their sterns were usually built up with a covered deck, whereas in British trawlers of the day the sterns were open and the big LL cable was stowed horizontally on the narrow side decks.

Then in 1941, a new class of small inshore sweepers was authorised; still called coastal minesweepers (and designated AMc) these craft were built of wood and carried an LL sweep, stowed aft on a reel which was mounted either horizontally, as in the trawlers, or vertically. The design was progressed very fast by a skilled Boston firm of naval architects and, in the end, 70 of these handy little craft were built. They were 97 feet long, and their diesel engines gave them a speed of 10 knots when running free.

The fourth significant advance made before Pearl Harbor was in the launching of the design of the new yard minesweeper. A special section is devoted to this important class in this chapter; suffice it to record here the important fact that, in part through an order placed by the Royal Navy under the Lend-Lease programme, in part as an urgent requirement for the US Navy itself, this class was

The two major war-built fleet sweeper classes; Foam, *of the American-built 'Raven' Class in the foreground, in service with the Royal Navy under Lend-Lease, and* Bramble *of the British 'Algerine' Class behind her. An American-built escort carrier of the Royal Navy is passing them.*

well under way by the time the end of 1941 came and, indeed, the YMS had appeared at sea well before that time.

The US Navy had the advantage of watching the mine warfare development in the European war for 2½ years before America joined in the global war. The Royal Navy gladly shared its experiences with its American ally, and this was of great help in saving time when hostilities opened for the Americans. Another advantage was the Lend-Lease programme itself, which gave an additional impetus to the speed-up of minesweeper design in America in that same period. The Royal Navy placed large orders for ships of the 'Raven' Class, and for the YMS, and both made a very important contribution to the minesweeping forces of the Royal Navy in the last two to three years of the war. Strangely, no similar orders were placed for ships of the 'Admirable' Class, though a large order was placed for the escort version (PCE) of the same design.

Atlantic Minefields

In 1942, during the period when German U-boats were doing well off the eastern seaboard of the United States, they laid a few minefields, as well as using their torpedoes to good effect. Some 340 mines were laid off American ports during this period and a total of ten ships, one a British naval trawler on loan to the US Navy at the time, were sunk or damaged by them.

The first fields, small ones, were laid off Delaware Bay and Chesapeake Bay, in June. The two fields contained only 30 mines between them but disrupted traffic considerably before the water was cleared. Three ships, two cargo and one tug, were sunk or damaged here, and a destroyer damaged by a near miss.

Then another U-boat laid two small fields further south, off Jacksonville in August, and off Charleston in September. These were all ground influence mines and were cleared by new sweepers of the AMc and YMS classes with little trouble. Further small fields were laid but, by then, the American sweepers were there in force, and little damage was done. No detailed records are to hand giving the allocation of minesweepers by command or fleet base, but it would appear that these fields tied up over 100 of the new small sweepers which were so badly needed in the Pacific.

Dan-buoys and Danlaying

Danlaying was by no means a new art in 1939. Fishermen for many years had used dan-buoys to mark the position of their nets and the naval versions, in the Allied navies, closely followed fishing practice.

Dans are made simply from a central pole, to the centre of which is fixed a watertight steel canister (fishing versions often used blocks of cork) so that the pole floats with half its length above the surface of the sea. A weight attached to the bottom keeps it upright and a wire runs from the middle of the pole down to a concrete sinker which moors the dan-buoy to the seabed. Some watertight pellets are added between the buoy and the mooring wire to prevent the latter dragging the dan under the water. It was important in waters with strong tidal streams, such as those around the coasts of England, to calculate the length of wire accurately so that high tides and strong currents did not drag the buoy under.

These dans were laid usually one mile apart, along the line of water swept and they carried brightly coloured flags to help the sweepers navigate very accurately by them. All the dans in one line would carry the same flag, with two flags on

Vallay, *a specialised dan-layer of the Royal Navy. Several of these were attached to each fleet sweeping flotilla; the dans can be seen stacked on the special deck aft, and the recovery davits can be seen forward of the bridge.*

each of the dans at the end. Usually, when fleet sweepers were clearing a field, two or three lines of dans would be left in cleared water, with danlayers picking these up at their leisure. But when clearance was being done under pressure, the danlayers needed to lay their buoys as fast as the sweepers could operate, and then pick them up at the rate of one buoy per minute, which called for real skill.

It seems strange that the Royal Navy in particular, which had been developing its wire sweeps actively for ten years before 1939, did not in the same period identify the need for a special and efficient danlaying vessel, especially to work with the new fleet sweepers. But, in the early years of the war, practically all danlaying was carried out by slow, elderly trawlers and drifters, which had been designed for handling fishing nets over the stern. They could not carry enough dans to service the fleet sweepers and they fell further and further behind as the laps progressed. The results were twofold. First, fleet sweepers and motor minesweepers had to carry dans, both laying and retrieving them themselves. Secondly, a number of the newer Admiralty-designed trawlers coming into service were converted as specialist danlayers.

For the first purpose 'Bangor' and 'Algerine' Class fleet sweepers received special fittings. Brackets to take snatch blocks for recovering dans were fitted to the outer forward edges of the Oerlikon gundecks below the bridge, and the normal minesweeping davits aft were used for laying the buoys, as our photographs show.

Another class of fleet sweeper adapted on occasion for danlaying by the Royal Navy was the excellent American-built BAM, or 'Catherine' Class. They had good speed and manoeuvreability, and some were fitted to carry 80 dans each. *Chamois* and *Chance* performed in this role at the Normandy invasion.

Similarly, seven 'Bangor' Class fleet sweepers, two manned by the Royal Navy and five by the Royal Canadian Navy, served at that same assault as danlayers; and the flotillas of BYMSs and MMSs present there allocated two out of every eight or ten vessels for the same duties.

Even in June 1944, of the 46 danlayers attached to the fleet minesweeping flotillas at the Normandy invasion, in addition to the nine fleet sweepers mentioned above, there were ten 'Isles' Class trawlers converted to danlayers, and 27 other trawlers which had similarly been converted—of which only two, of the 'Round Table' Class had been built during the war.

However, war-built trawlers were being converted as fast as the overburdened repair yards and shipyards would allow. Eight of the 'Tree' Class were found to be suitable; the 'Isles' Class trawler *Sheppey* was converted at Gibraltar, and a further 25, of which only eight were coal-burners, were converted on the stocks.

In these conversions, speed in laying and recovering dans, to keep up with the increasingly fast fleet sweepers, was the overriding requirement. Whilst the 'Isles' Class were picked as the largest and fastest of the naval trawlers building, even they were not quite fast enough for this job, hence the great effort put into improving their handling gear. The target performance was that a dan-buoy, complete with a 50-fathom mooring and its sinker, should be recovered in one minute. As a further measure of the efficiency required, the Admiralty laid down that where unconverted trawlers were used as danlayers from the end of 1943, then three should be allocated to each fleet sweeping flotilla, in place of the normal two.

Far East and postwar mine clearance

The Admiralty was planning and estimating the numbers of ships needed for these two big tasks as early as 1943. In November of that year the estimate of postwar requirements in home waters alone was for 88 danlayers, plus a further 18 ships, all oil-burning, for the two Far East fleets—a total of 106 efficient danlayers. By September 1944, however, this figure was revised downwards to 88, to include a reduced requirement of eight ships for the Far East fleets, as the war there moved at a faster pace. This was later reduced again to 48 ships for home waters—probably related to the reduced availability of fleet sweepers, as the postwar manning situation threw up problems during the demobilisation of wartime personnel.

The US Navy never did produce a specialised danlaying vessel. The efficient yard motor minesweepers were equipped to carry dans of a very similar design to that used by the Royal Navy except that the central canister was of conical shape and flags were not always carried. Fleet minesweepers (AMs) also carried dans, and used them as necessary. The sinkers hooked to the edges of the upper decks figure prominently in some of the photographs of these ships.

One very different requirement that arose during the war was the laying of lighted dan-buoys to mark the channels being swept during the assault operations which, as often as not, entailed an arrival for the landing craft at dawn.

126-foot motor minesweepers as danlayers

Some of this class were refitted for danlaying duties; it is believed that MMSs 1079, 1087, and 1088 were altered in this way, whilst still under construction. As with the 'Isles' Class, all minesweeping and related gear was removed. A new dan-handling deck was built aft to the same level as the forecastle, and the funnel was heightened by five feet. The outfit of dan-buoys was fixed at 43; 16 rigged buoys were stowed on the starboard side, nine on the port side and an additional 18 unrigged dans were stowed on the centre line. Forward, samson posts were fitted, as in the 'Isles' Class conversions, and all the gear for handling dans quickly was included—the whole ship was reorganised for this essential duty.

Destroyer mineweepers (DMSs—USN)

This was essentially an American ship type. The Royal Navy did fit high-speed destroyer sweeps on the sterns of a number of fleet destroyers before the war, but they were not used to any great effect during the war—except in the Mediterranean, where destroyers swept ahead of the famous convoys which relieved the island of Malta.

However, in the US Navy, especially for operations in the great open spaces of the Pacific, destroyer minesweepers were much in use, and did a great job. In the assault operations, it was quite usual for destroyer sweepers with their high speed to sweep in ahead of the assault, with the smaller fleet sweepers following up with the main force.

American destroyer minesweepers took two forms. The first was a group of 18 old 'four-piper' destroyers from World War 1, which were given Oropesa sweeps at their stern; they performed with great distinction in the early years.

Then, when new destroyers were appearing from the builders' yards in great numbers in 1943 and 1944, 24 units of the new 'Livermore' Class were converted to destroyer minesweepers. In addition to the Oropesa sweep aft (for which they sacrificed their after gun mounting) they were fitted with the LL magnetic sweep aft. The SA acoustic sweep was also fitted, with a towed box handled by a short derrick amidships.

Minesweeping corvettes and frigates (RN)

When the 'Flower' Class corvettes appeared the mine warfare campaign was at a crisis point and, although the North Atlantic battle was also critical, the shortage of fleet minesweepers made it essential that at least some of these new ships, of ideal size, should be made available for sweeping. So it was not surprising that in 1940, 24 of the 'Flower' Class were fitted with Oropesa sweeps and a further eight (later 16) of the class were fitted with LL sweeps. In 1941, those ships fitted with the LL sweep also received the SA sweep, with the version mounted on a boom hinged over the bows.

Smith's Dock, the lead yard for the class, mocked up the first Oropesa version in September 1940, but there were many difficulties in fitting all the gear into the space available. The number of depth charges had to be severely restricted (a maximum of ten), and the displacement went up significantly.

The 'Flower' Class corvette Hyderabad *in April 1942, with SA sweep on a boom over her bow, and LL reel mounted on her stern, forward of the depth charge racks.*

It seems doubtful whether this class of corvette was used for minesweeping in UK waters, but a group of them went to the Mediterranean in 1941 and performed valiant work. Both Oropesa and LL/SA versions were at work there around Greece, Crete, Malta, and the North African coast. Later, a group went on to join the East Indies Fleet.

When the first of the 'River' Class frigates appeared, the mining crisis was still dominant and the early units were fitted with a full range of Oropesa, LL, and SA sweeps. In February 1942, they were being referred to as 'fast corvette minesweepers'. Their maximum speed through the water, with Oropesa sweeps out, was 11 knots, and their SA gear was mounted internally. It seems doubtful if this class were ever used operationally as sweepers, since their presence was so urgently required in the North Atlantic.

Yard motor minesweepers (YMS—USN)

All units of this class were built in the USA. Of the 550 ships completed as minesweepers 234 were transferred to other navies.

The design and rapid construction of this impressive class of coastal minesweeper (which served all over the world) is one of the most exciting even among all the wartime naval construction programmes of the Allies. There were two major differences between mine warfare in World War 1 and the vastly more complicated version of 1939–45. One was the dramatic effect of the use of ground mines of all types; the other was the new requirement for efficient assault sweeping as the spearhead of the many landings both in Europe and in the Pacific.

These new and demanding sweeping requirements threw up the need for a new type of sweeper, first for the Royal Navy, then for the US Navy as well. In both, they were called motor minesweepers, but in America, the word 'yard' was added in front, as these ships were at first classified among the craft attached to a navy yard, or base, and not expected to go beyond the adjacent waters. The same basic classification had been made by the Royal Navy, with the roles of the YMS and the MMS being very similar.

Left *YMS 74, with a double Oropesa sweep streamed. The wires can be seen running out over her stern, and the extended Oerlikon platforms of this class, just aft of the bridge, can be clearly seen.*

Right *The American YMS20 streaming her starboard Oropesa sweep. The float is handled by the midships derrick, and the otter by the stern davits.*

However, the original classification was soon overtaken, for the American YMS, like its counterpart in the Royal Navy, eventually made long ocean voyages and served in all parts of the world, wherever its special sweeping qualities could be put to good use. The long ocean voyages meant special preparation—extra fuel stowed in drums on deck and larger craft to escort them on long exposed passages. The YMS suffered badly in the violent typhoons of the western Pacific in 1944 and 1945—this comes out clearly in the long list of war losses of this class—equal numbers by enemy action, and by storm, grounding, or collision.

The YMS (again like the MMS) was of all-wooden construction, to reduce its sensitivity to magnetic mines, and so, by virtue of its size and building materials, it was usually built by yacht yards rather than by the larger shipbuilders.

It was designed by Henry B. Nevins, Inc, of City Island, New York, with additional ideas from two other design sources. As with the destroyer escorts, the Royal Navy's request for 150 of the YMS class under Lend-Lease gave an extra spur to the effort already under way by the US Navy, which had been following closely the mine warfare developments in Europe. So it was that a large number of YMSs were laid down before Pearl Harbor, in addition to those units being built for the Royal Navy.

All of the YMSs were built in America, and all to the same basic design. YMSs 1 to 134 had two thin funnels, YMSs 135 to 445 and the last two, YMSs 480 to 481, had one large and distinctive funnel; and YMSs 446 to 479 (a relatively small group) had no funnel at all.

Originally, they were fitted for both wire (Oropesa) and influence sweeping; it is not clear whether this configuration was maintained in all YMSs during the war or if some had their wire sweep gear removed, leaving them free to concentrate on magnetic and acoustic sweeping. But big winches for wire sweeping were fitted aft, together with small davits on the quarters for handling the floats, otters, and kites. The floats were sometimes stowed aft, sometimes amidships, under the heavy two-armed derrick fitted aft of the funnel. The latter was used both to handle the boat and the towed-box version of the acoustic sweep, fitted in later versions of the YMS. The big LL reel was fitted just forward of the winch and, in the earlier units, the acoustic box was fitted on a boom over the bow. Both this gear, and the later towed-box SA gear, was very similar to that in use by the Royal Navy. The only significant differences were in the design of the wire winch and in the davit lay-out.

The YMSs served the US Navy magnificently, and especially in the Pacific, where their growing numbers were essential to the increasing tempo of the assault landings on the islands, and to the clearance sweeping of liberated waters. The saga of that campaign is liberally filled with stories of these gallant small sweepers.

The YMS design itself was an impressive one and it is not surprising that it was developed as the design for the postwar coastal minesweeper and minehunter and built in a number of forms for many Allied navies all over the world in the two decades after 1945. The YMSs lasted for many years after the war, though in gradually reducing numbers; but they lasted longer than the Royal Navy's MMSs. Those YMSs still in service in 1947 were reclassified as AMSs (auxiliary minesweepers); in this role, they were re-rated as mine warfare ships. In 1955, their type symbol was changed again, to MSC(O), Minesweeper, Coastal (Old), and again to MSCO in 1967. They provided the bulk of the minesweeping capability for the Korean War, and the last YMS (327, later known as RUFF, MSCO–54) was only struck from the US Navy List in November 1969.

The YMS construction programme

Covering 561 units in all, this was one of the largest single classes of warship produced during World War 2.

Speed of completions

	1942	1943	1944	1945
1st quarter	48	41	13	9
2nd quarter	54	63	12	1
3rd quarter	54	102	23	1
4th quarter	54	59	27	0
Totals	210	265	75	11

The large number of completions in 1942 is especially interesting. A total of 116 YMSs were laid down before Pearl Harbor and this number does not include any of the units being built for the Royal Navy under Lend-Lease.

The average quarterly production was remarkably consistent; the third quarter of 1943 showed a clear peak figure, and there was an equally clear rundown during 1944. If conclusions may be drawn from these figures, then they are that the US Navy's estimated requirement for YMSs was 400 units, plus the Royal Navy's need for 150 units (note that the latter was the only case within the Lend-Lease programme where the original number requested was actually delivered). Note, too, that there were no cancellations—the only major American-built class where this did not occur.

The builders of the class

No less than 35 yards were involved in building YMSs. Of these, 12 were on the eastern seaboard, 19 on the West Coast, and four on the Great Lakes. Thirteen of the yards had one order only, each for a number of units. Nor was any concentration of repeat orders focused on the yards with the fastest building times; they were pretty evenly spread. To draw an arbitrary distinction, some 150 units were ordered from fast yards, some 200 from yards with average building times, and some 125 from the slower yards. Probably the reason for this spread was that the YMSs were ordered from the smaller yards, which were not so easily adapted for mass-production methods. Indeed, we can show that the larger the number of building berths, the greater the problems! The relatively small size of the ships and the fact that they were of all-wooden construction also indicated a large number of small yards used to constructing wooden yachts in peacetime.

All this makes a very interesting comparison with the Royal Navy's MMS building programme; but in the UK there were extra problems to cope with as well, such as air raids on the yards.

The fastest yards

Henry B. Nevins, Inc of City Island, New York, built 24 units. The building times were drastically cut during the construction of a batch of 11 YMSs, from some 13 months each to three months and 18 days, the fastest time achieved by any of the builders. Excellent times were also recorded by this yard with a second batch, especially in the peak period of mid-1943. The yard seems to have had three berths.

Greenport Basin and Construction Co of Long Island, New York, built 38 units. They, too, cut building times successfully, their fastest being four months and 19 days, with a low average time in addition—an impressive record. And this with just two building berths.

Associated shipbuilders of Seattle, Washington, built only one order, of ten ships, but had a very fast average time; the second fastest of the lot—three months and 30 days.

Seattle Shipbuilding and Drydock Co also in Seattle, achieved fast building times with only one order.

Slower yards, by comparison, were taking from ten to 16 months, with many units completed in the 12–14 months range. The differences were considerable.

The building times for the YMS were remarkably similar to those for the larger DEs, both for the fastest times and for the slower ones. At first sight this seems remarkable for ships that were less than half the size of the escorts, but the difference lies in the mass-production methods developed for the latter, which were not suitable for the YMS construction programme.

Surprisingly, the number of berths in a yard was not a significant factor in achieving a high production rate. Indeed, if anything, the reverse seemed to be the case! For example, one East Coast yard had no fewer than ten building berths and, on July 22 1941, that yard laid down ten YMSs on the same day! But then things got difficult, probably on delivery of components, and completion dates were progressively three weeks later between units, starting at 11 months nine days and finishing at 19 months. The same thing happened during the construction of a second batch at the same yard, doubtless for reasons entirely outside its control. Another yard seems to have had eight building berths but had only one order for the YMS Class, and was not able to work fast. These comments are in no way intended to be critical. Rather do they show that fast building times depended on the achievements of individual yards, rather than on mass-production assembly lines, when building this specialised type of vessel. The whole programme was carried out under great pressure, and the results by any standards were most remarkable.

Slowdown in 1944

The destroyer escort and patrol craft escort building programmes slowed down drastically in 1944, when it was thought that the Battle of the Atlantic was under control. A similar trend can be seen with the later units to be completed within the YMS building programme. Presumably, as the total of 400 US Navy YMSs was approached, it was felt that the number of yard minesweepers reaching the battle areas was confirming that the original target figure was correct, and that emphasis could be switched to the pressing needs for ever more landing craft. One Florida yard, for example, achieved excellent building times for YMS units completing in 1942 and 1943; but a further batch, completing in the second half of 1944 averaged $1\frac{1}{2}$ to 2 times the building period of the earlier units.

Transfers

The wartime total of transfers was 234 units. There were three major groups of YMSs transferred under Lend-Lease to Allied nations during the war years. Some 150 were transferred on completion to the Royal Navy, where they made a very significant contribution to the sweeping forces. Note that 80 of these were specially numbered BYMS 1–80, and did not carry any YMS numbers before transfer. All of these units were picked up at their builders' yards by Royal Navy crews.

Thirty YMSs were transferred to France—25 of them with YMS numbers under YMS 100. All had substantial US Navy war service prior to transfer; seven hoisted the French Tricolour in March 1944, and the other 23 in October 1944. All were sold outright to France in 1949, except one, which was sunk off Marseilles in October 1944.

Forty-two YMSs were transferred to Russia; again, all had substantial US Navy war service prior to transfer—six in March/April 1945, the rest between May and September 1945.

Then there were smaller numbers transferred to other Allied nations—four to Greece, for example, after Royal Navy war service. They were turned over in the Mediterranean to their new crews. Eight went to Norway in 1945, and all these transferred ships made a good contribution to mine clearance operations after the war.

Postwar transfers

These totalled 34. Nine were transferred in Europe for mine clearance—two to Holland, four more to Greece, and three to Turkey. Six of these had seen war service with the Royal Navy. Twenty-three were transferred in the Pacific—11 to Korea, seven to Japan, and others to Thailand, the Philippines, and Nationalist China.

War losses of US YMSs (in numerical sequence)

By enemy action

YMS 19	Sunk by mine	Sept 24 1944	Caroline Islands
YMS 21	Sunk by mine	Sept 1 1944	Southern France
YMS 24	Sunk by mine	Aug 16 1944	Southern France
YMS 30	Sunk by mine	Jan 25 1944	Off Anzio

YMS 39	Sunk by mine	June 26 1945	Balikpapan area
YMS 48	Shore batteries	Feb 14 1945	Corregidor
YMS 50	Shore batteries	June 18 1945	Balikpapan area
YMS 71	Sunk by mine	April 3 1945	Off Borneo
YMS 84	Sunk by mine	July 9 1945	Balikpapan area
YMS 103	Beached after mining	April 7 1945	Okinawa
YMS 304	Sunk by mine	July 30 1944	Normandy
YMS 350	Sunk by mine	July 2 1944	Normandy
YMS 365	Sunk by mine	June 26 1945	Balikpapan area
YMS 378	Written off after mining	July 30 1945	Normandy
YMS 385	Sunk by mine	October 1 1944	Caroline Islands
YMS 481	Shore batteries	May 2 1945	Tarakan

By storm, grounding, or collision

YMS 14	Sunk by collision	Jan 11 1945	Boston Harbour
YMS 70	Sunk in storm	Oct 17 1944	Off Leyte
YMS 98	Sunk in typhoon	Sept 16 1945	Okinawa
YMS 127	Lost by grounding	Jan 16 1944	Aleutian Islands
YMS 133	Foundered	Feb 21 1943	Off Oregon
YMS 146	Sunk in typhoon	Oct 9 1945	Okinawa
YMS 151	Sunk in typhoon	Oct 9 1945	Okinawa
YMS 275	Sunk in typhoon	Oct 9 1945	Okinawa
YMS 341	Sunk in typhoon	Sept 16 1945	Okinawa
YMS 383	Sunk in typhoon	Oct 9 1945	Okinawa
YMS 409	Foundered in hurricane	Sept 12 1944	Eastern seaboard
YMS 421	Sunk in typhoon	Sept 16 1945	Okinawa
YMS 424	Sunk in typhoon	Oct 9 1945	Okinawa
YMS 454	Sunk in typhoon	Oct 9 1945	Okinawa
YMS 472	Sunk in typhoon	Oct 16 1945	Okinawa
YMS 478	Lost by grounding	Oct 8 1945	Japan

In addition, two YMSs were lost in the Korean War in 1950–51.

YMSs lost while serving with the Royal Navy
BYMS Nos. 2019, 2022, 2030, 2055, 2077, 2255.
Details appear under the Royal Navy Section.

YMSs lost while serving with other Allied navies

YMS 77	Sunk by mine	French	Oct 25 1944	Marseilles
YMS 190	Sunk by mine	Greek	Apr 15 1944	Cape Turbo
YMS 382	Sunk by U-boat	Norwegian	May 7 1945	Cherbourg
GYMS 2191	Sunk by mine	Greek (Ex-RN BYMS 2191)	June 5 1945	Aegean
GYMS 2074	Sunk by mine	Greek (ex-RN BYMS 2074)	Date not identified	Aegean

Small sweepers

There were a number of small, mostly wooden-built craft which served usefully as inshore sweepers or danlaying vessels during the war. This is not easy to identify from contemporary records of these classes (for instance, of the Royal Navy's total of some 700 112-foot 'B' Class motor launches, some 130 were fitted with light rope Oropesa sweeps by the end of the war, and a few others with magnetic sweeps).

Royal Navy

112-foot motor launches appear in this book often. They were first used effectively as Oropesa sweepers at Malta in 1942 and 1943. Later, they served as skim sweepers ahead of the leading fleet sweeper in assault operations—at Normandy, they performed this successful role for both British and American fleet sweepers. Some were also fitted as influence sweepers and did some good magnetic sweeping in both home and Mediterranean waters. One such ML was sunk right in the entrance to Ostend Harbour, while sweeping in the small assault force in September 1944.

Harbour defence motor launches were smaller than the 'B' type, only 72 feet overall. Some were fitted with light Oropesa sweeps, and were used in harbour approaches, in shallow waters and in rivers after capture. A light steel gallows can be seen fitted over the stern of some of these boats in photographs.

Motor fishing vessels were used for similar purposes, but were more heavily built, and whereas the MLs of both types were largely constructed by peacetime yacht builders, the MFVs were built by the traditional yards specialising in fishing craft. They were usually fitted with influence sweeps, and not least for freshwater sweeping; but their greater draught made them less suitable for this work. They, too, were built in great numbers, but the total actually fitted for sweeping is not known.

US Navy

The largest of the smaller sweepers were the PCSs, converted while under construction from the steel 170-foot Patrol Craft. Seventeen of these craft were pressed into service as sweepers, in the critical period before the YMSs rolled out of the shipyards in great numbers; later, they were re-converted to their Patrol Craft role. A large LL sweep was mounted aft, though the narrow breadth of these ships made them less than ideal for the purpose. But they had a free speed of 20 knots and carried a good armament of one 3-inch gun, plus five 20 mm Oerlikons.

Submarine chasers were the American equivalent of the British motor launches. Of 111 feet overall they, too, had a speed of 20 knots free, and were built of wood. Statistics are not available, but it is probable that a number of these boats were fitted with light Oropesa sweeps for assault work.

Chapter 5

First assaults, and bigger sweeper fleets, 1943

Of all the war years, this was the least worrying for the sweepers—though clearance and losses continued without a break. But no new types of mines appeared, though variations on the magnetics and acoustics had to be watched for, and preparations were going full blast to counter the pressure mine, whenever it might appear.

New sweepers were commissioning in great numbers, and it is interesting to look at the Admiralty's requirement for fleet sweepers, defined in this year, and compare it with the prewar position. It bears little relation to the 1939 availability.

Royal Navy fleet sweeper requirements

In March 1943 the Admiralty defined the role of the fleet sweepers. It was, first, to have sufficient ships to deal with the enemy's offensive minefields, day by day and, secondly, to have sufficient ships to break through the enemy's defensive minefields, in any assault area in Europe.

It was decided that the new specialisations in equipment on the North Atlantic station meant that fleet sweepers, fitted for anti-submarine work, could supplement the escorts but not vice versa. So the numbers of the new 'Algerine' Class fleet sweepers building were increased, in part by ordering more units of the class from Canada, while the United Kingdom shipyards specialised in building the new frigates. This brought two extra flotillas of 'Algerines' into commission before the invasion of Normandy.

The number of fleet sweepers required for 1944 and 1945, with the assaults really getting under way, was as follows:

Home waters

Station	**Ships needed**		
Home Fleet	3 flotillas	=	24 ships
North Russia	2	=	16 ships
UK West Coast	1	=	8 ships
UK East Coast	5	=	40 ships
English Channel	2	=	16 ships
Total, home waters	13 flotillas	=	104 ships

Although all fleet sweepers needed for assaults in Europe would have to be taken from these numbers it was thought that even if all these ships were built under pressure the manpower would not be available to commission them, and so the numbers of ships required were reduced to seven flotillas (56 ships).

Abroad

Station	Ships needed		
Mediterranean	4 flotillas	=	32 ships
Canada	6	=	48 ships (RCN)
Indian Ocean	$2\frac{1}{4}$	=	18 ships (RIN)
Eastern Fleet	4	=	32 ships (incl RAN)
Pacific Ocean	6	=	48 ships (incl RAN)
Total, abroad	$22\frac{1}{4}$ flotillas	=	178 ships

Of these totals, it was calculated that the Royal Navy would itself need to provide 176 fleet sweepers and, at the beginning of 1943, only 89 were in service. With a maximum building programme, and allowing ten per cent for casualties in any one year, progress needed to fulfil that need looked like this:

	Fleet sweepers to be available
July 1943	113
January 1944	137
July 1944	154
January 1945	162

These still did not meet the full need, but used the Commonwealth shipbuilding capacity to its best advantage. In 1944 and 1945 old capital ships and cruisers were being paid off into reserve in the United Kingdom to provide more skilled manpower for the new fleet sweepers, as well as for the new escort frigates and corvettes.

However, figures alone do not give the true picture of the real turn-round in the availability of fleet sweepers. Apart from the numbers growing in parallel of new MMSs and BYMSs, the Royal Navy in January 1943 had the following fleet sweeper flotillas in service:

Home waters	Mediterranean	In reserve
2 'Algerine'	2 'Flower' (corvettes)	2 'Flower' (corvettes)
2 'Bangor'	1 'Algerine'	1 'River' (frigate)
1 'Town'	3 'Bangor'	
2 'Halcyon'	1 'Town'	

United Kingdom home waters

The number of ground mines swept was only half that of the previous year and a quarter of the total in 1941. The moored mine total also fell by half, and the number of ships lost to mines was down to just ten in the year, compared with 68 in 1940.

From the beginning of the year, the new and highly effective BYMSs were crossing the Atlantic in numbers, from their builders' yards in North America, while the MMSs, now at sea in large quantities, were influence sweeping in these waters. More of the older trawlers were being phased out of service and most of the skids, which had been kept in reserve for a couple of years, were finally towed away from the crowded sweeping ports to an honoured retirement.

A motor launch flotilla, fitted with Oropesa sweeps, was now operating from Dover and cut 72 moored mines off the port in just a few days. But a new and improved type of German E-boat was also in service, each carrying six ground

BYMS 172, a later version, still flying the stars and stripes on trials before being taken over by a Royal Navy crew.

mines, and the sweepers were worried about them laying a field just ahead of a coastal convoy. Diversion channels were charted and kept swept, so that the convoys could be quickly kept clear if this danger arose and, in fact, the convoys used this ruse on several occasions at night.

At Dover, shelling from the big German guns at Cap Gris Nez near Calais was always a danger. MMS 287, sweeping with two others, was badly damaged in this way and, in September, the Ninth and Eighteenth Flotillas of fleet sweepers were bombarded heavily while clearance sweeping in the Dover Straits. *Hydra* was straddled and hit just outside Dover breakwater, and severely damaged. *Qualicum* had 18 near misses. The sweepers retired hastily under cover of smoke laid by their attendant motor launches.

The year ended on yet another encouraging note in these hotly contested waters—the arrival from America of *Gazelle*, the first BAM, or 'Catherine' Class fleet sweeper, built under the Lend-Lease programme, and a fitting partner for the new 'Algerine' Class built in the United Kingdom and Canada.

The Mediterranean—the assaults begin

The year started badly for British sweepers here—several nasty losses were encountered. First, the 'Algerine' *Alarm*, of the Twelfth Flotilla, was bombed off Bone in North Africa in January, her back was broken and she had to be beached. Then *Speedy* was mined outside Malta's Grand Harbour, MMS 89 was bagged by a moored mine off Bizerta in May, to be followed by HDML 1154 off the same port. In the same month, the 'Algerine' *Fantome*, also of the Twelfth Flotilla, lost her stern to a mine during the passage of the first convoy from Alexandria through to Gibraltar, after the relief of Malta.

Operation 'Husky'

Soon, in July, the great assault on Sicily was planned and then launched from Malta. Once more the Grand Harbour was packed with warships and merchant ships, and this time the Stars and Stripes flew everywhere among the White,

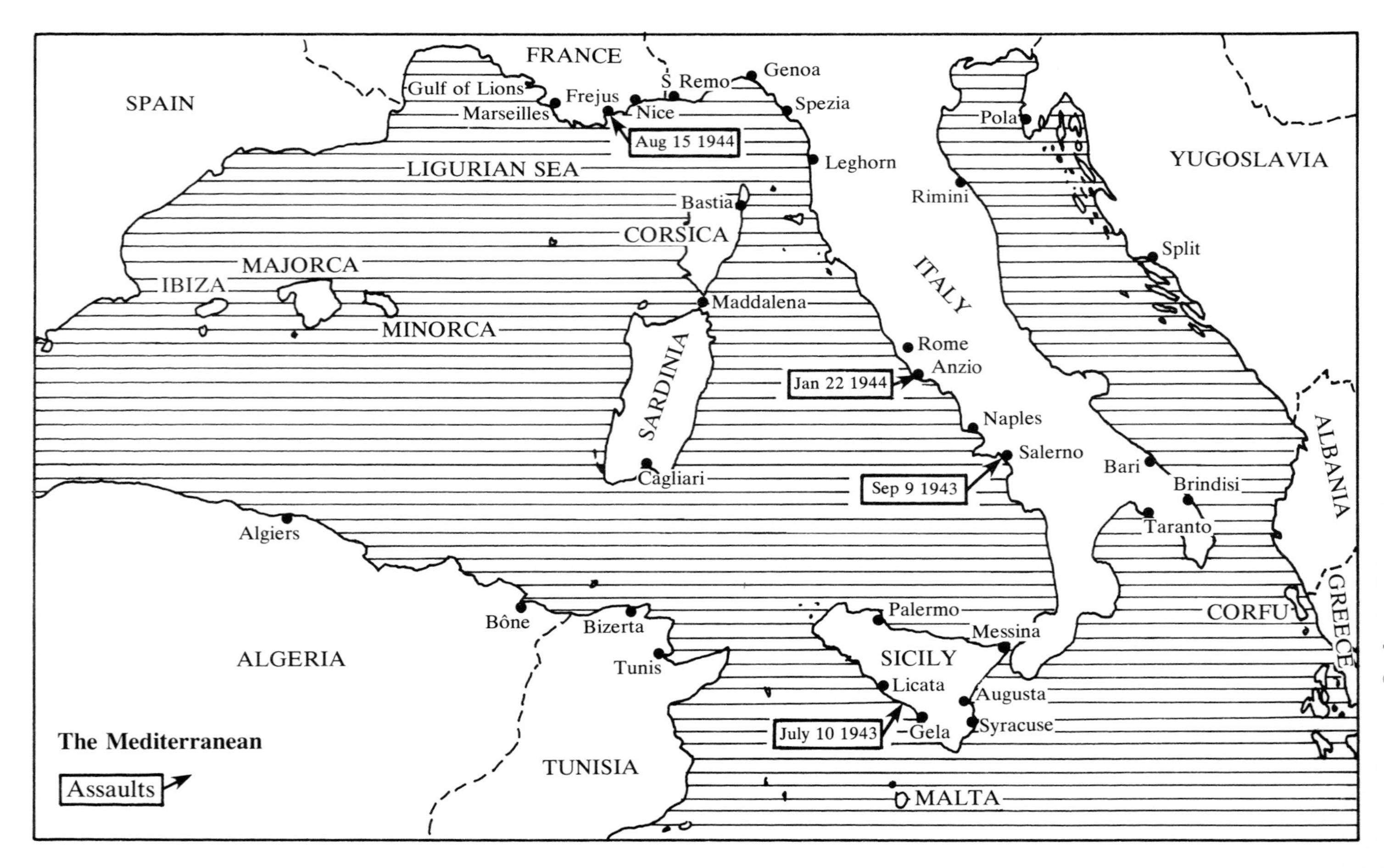

The Mediterranean

Blue, and Red Ensigns. Many of the strange new landing craft were there—LSTs, LCTs and LCIs. The minesweepers were there, of course, in full force, British and American fleet sweepers and coastal sweepers in profusion.

From the sweeping point of view, the invasion went very well. The sweepers cleared the approach channels, in rough seas, and then led the assault landing craft in towards their beaches. Then they spread out and cleared lateral transport anchorages and fire support areas parallel with the beaches. Next, they continued non-stop maintenance sweeping and, in addition, carried out anti-submarine and anti-aircraft patrols to seaward.

The sweeper fleet present at the landings was an indication of the build-up of ships now coming from the vast new construction programmes. Although this was only the first of the assault operations, the British sweepers numbered eight of the fleet classes, ten MMSs, and six BYMSs; while the American force included ten fleet sweepers (AMs) and 25 of the yard motor minesweepers.

Several of the sweepers received minor damage in air attacks, but only the American AM *Sentinel* was sunk, 15 miles off Licata, by dive-bombing. In fierce seas, her survivors were safely taken off by two small anti-submarine patrol boats.

The only enemy minefield of any size was found off Empedocle, and cleared in short order by the American sweepers. Twenty-one mines were lifted on the first day, under accurate enemy shellfire from the shore, but the American AM *Staff* ran over a mine and was towed in severely damaged. In all, 63 mines were swept by the Americans in this operation.

Then clearance sweeping was completed all round the island and, at last, the main channel through the Mediterranean towards the North African shore could be cleared for the through convoys. Three fleet sweeping flotillas from Malta, with numerous MMSs, trawlers, and MLs, took part in this intensive operation, clearing a swept channel no less than 200 miles long and a broad two miles wide. Over 200 moored mines were cut.

BYMS 2022 off Cotrone, in the Mediterranean, in November 1943. She was working as a dan-layer at the time, two dans are ready for dropping over her stern, and others are stacked ready for use amidships.

Two American YMSs lifting mines off Salerno, in 1943. These excellent small sweepers performed great service in all major theatres of war.

Operation 'Avalanche'

The assault on Salerno, on the Italian mainland, came in September. The sweeper fleet was impressive again—15 British fleet sweepers, plus nine American AMs, and 23 American YMSs. The sweeping plan was ambitious, and the sweepers worked hard and successfully; but this was still in the early days of assaults, and large warships, transports, and landing craft were still happily straying out of the newly swept channels, both fouling and cutting the sweeps. The assault went in successfully; in the British sector, 135 mines were swept near the beaches in a single day. Seven American YMSs were working in this area; an interesting comparison with the Normandy landings, where so many British sweepers were employed in the American sectors. A further 124 mines, all British-laid, were cleared from a field off the magical isle of Capri. In a burst of enthusiasm, the British Thirteenth Flotilla of fleet sweepers cleared a way into the Bay of Salerno, but they were a bit ahead of the game, and got shelled for their trouble. The leading British sweepers laid lighted dans astern of them, and markboats were prominent, with lights flashing. Many mines were caught in the sweeps, but were successfully cleared, and the American YMSs worked in happy partnership with the British MLs from the famous Malta sweeping flotilla in skim-sweeping ahead of the leading fleet sweepers. This was the first time that this technique had been used in a major assault.

In the American sector, there were many snagged and broken sweeps, but the YMSs alone cut 52 mines, and the AMs were sweeping continuously for 31 hours. One American AM was lost, *Skill*, torpedoed on September 25, and she sank very quickly.

The year in the Mediterranean closed with continuing sweeper losses, to underline the never-ending nature of the sweepers' task. *Hebe*, of the 'Halcyon' Class, was mined and sunk off the entrance to Bari in November, while *Felix-*

stowe, a 'Bangor' Class fleet sweeper, was sunk in the approaches to Maddalena—the third sweeper to be lost in an area which was to claim more victims yet.

Pressure mines and sweeps

The pressure mines (code-named 'oysters' by the Allies) were first used by the Germans on the Normandy beach-head in June 1944; but the Royal Navy had made intensive efforts in the previous year to devise a sweep against such a mine. Pressure mines were the last major innovation of World War 2, and they remained virtually unsweepable right up to the end of the war. There were other ingenious devices produced, true—200-day arming clocks, for instance—but it was the pressure mine which constituted a major threat during the last two years of the European war. Only recently have the security wraps come off this fascinating story.

Briefly, these mines reacted to the decrease in water pressure under a ship's hull, as she passed by—an effect which becomes more pronounced if the ship is travelling fast or in shallow water. The mine incorporated a suction device which was closed by this change in water pressure as the ship passed overhead thus firing the explosive charge. 'Oysters' were often combined with magnetic or acoustic units so that the closing of one of them needed the reaction of the second unit before the mine could fire.

Early sweeps

The main threat was to Allied shipping in Europe, and the Royal Navy worked hard to produce an antidote. The idea was not new—in 1917 such a mine had been proved possible and both Germans and British put a big effort into producing one from 1939 onwards. Within a year, both sides had produced a workable mine; but neither side used it, since it was an unwritten law (perhaps of self-survival) that a mine was not laid before a sweep effective against it had been found by the layer.

Wavemaking sweeps were the first British effort. Fast MTBs, 115 feet long, were used in a number of experiments, notably at Portland. They ran at full speed, three abreast, over a simulated mine, but the suction period as they passed by could not be maintained for long enough to fire the mine. Later, in 1944, when the 'oysters' actually appeared, further experiments were made with wavemaking sweeps. Some of the latest and fastest MTBs, and even the one flotilla of the larger but very fast steam gun boats, were used, the latter carrying specially adapted magnetic and acoustic sweeps to catch the combined mines. Sweeping in line abreast at 20 knots, they swept a path 300 yards wide and the outer boats laid dan-buoys to mark the swept channel. But, as far as is known, they never swept a mine.

Another idea, used by both the British and Americans, was to have a destroyer steam at 30 knots towards the suspected minefield, then suddenly turn off, leaving her wave pattern to roll on over the field and, it was hoped, explode the mines. But even with LL sweepers stationed outside the field to catch the magnetic side, this was thought to be too dangerous. In 1945, however, this idea was tried yet again. The new fleet destroyer *Wakeful* was used in extensive experiments. Other ship types tested in this role at various times were 'Captain' Class frigates and 'Catherine' Class fleet minesweepers.

Counter-mining from a distance seemed a good idea, and again, both the

British and Americans tried hard. British experiments showed that 300 depth charges per mile would be needed to clear a channel 100 yards wide, so this idea was abandoned. American attempts used charges of no less than ten tons of explosive each. These cleared an area with a radius of 600 yards, but this idea was not taken further—perhaps because of the effect on the nerves of the sweepers! The Germans did, however, have some success later against British-laid pressure mines, by dropping depth charges outside the minefields.

Against acoustic mines, the British had successfully adapted ordinary road drills and metal poles; now against pressure mines they tried an ordinary barrage balloon (code-named 'Beta'), as used against low flying bombers. This was the first sweep designed to create a suction in the water similar to that of a big ship passing by. The balloon was filled and towed through the water at 6 knots, giving it the same resistance as a wind speed of 50 miles per hour. The towing strains were great, the balloon oscillated violently in the water, and the attempt was abandoned.

Then a trawler's rope cone net was tried; it had a length of 140 feet, and a towing strain of ten tons, but the 'pressure signature' which it gave was too low. Japanese sweepers did, however, use sleeve trawls with some success in the Pacific, against American-laid mines, in mid-1945 and in postwar clearance.

Code-named 'Delta' (for 'Displacement sweep') the most promising development, in 1942 and 1943, seemed to be the construction of two special 'dumb', or engineless, barges, designed to create a suction simulating the passing of a large ship, with minimum resistance to towing, and maximum resistance to the effect of exploding mines. Two such barges were ordered for the Royal Navy in 1942, one from Denny on the Clyde and one from Swan Hunter on the Tyne. Built under the greatest secrecy, they were officially known as 'Stirling Craft', then as 'Algerine' Class fleet minesweepers, being given the names *Cyrus* and *Cybele*, and referred to as 'His Majesty's Ships'.

They were built of steel, with a displacement of 4,000 tons; they had a length of 361 feet, a breadth of 65 feet, and a draught of 25 feet. Their dimensions were

One of the 'Stirling Craft' on the stocks on the Clyde, in Scotland, in 1943. While records of these two unique sweepers against pressure mines are thin, this shot gives some idea of their unique construction.

fairly close to the war-built merchant ships which they were largely designed to protect.

No drawings remain of these remarkable ships, and the photograph in this book is thought to be unique; it was identified in the Denny collection at the National Maritime Museum in London by David Lyon. These barges appear to have had a lower section of rectangular shape, open in the centre, with strong cross supports. They had normal vertical outboard sides and a light bottom plating. Above this box, which probably had a bow and stern of fairly normal shape, were five long, slim hulls joined together by strong girders. The theory apparently was that when a mine exploded beneath the hull, the light bottom plating would collapse and the hull would sink until the buoyancy hulls took the strain; the barge would then be towed back to port for repairs to the bottom plating.

Cyrus, completed in the summer of 1943, underwent 'first of class' trials in the Clyde. An early problem was encountered in the tugs needed to tow the barge at operational speed. The largest war-built rescue tugs, of the 'Bustler' Class (205 feet long), and the 'Assurance' Class (157 feet long) were used, but even at low speeds through the water the pressure disturbance under the tugs was nearly as great as under the barge itself, thus putting them in great danger.

The experimental explosions were also disappointing; the first simulated mine strike against *Cyrus*, in August 1943, badly damaged her stern: *Cybele* was completed in January 1944, but an explosion under her caused equally heavy damage. *Cybele* was then tested to destruction, but *Cyrus* was laid up in damaged condition in the Clyde and, after that time, they were referred to in naval records as 'HM Hulk *Cyrus*', and so on.

'Egg Crates'

When the 'oysters' appeared in 1944, urgent arrangements were put in hand to construct more barges to operate as displacement sweeps. The resulting barges did not operate until 1945 at sea, and then only spasmodically, but their story is told here, to keep the picture intact.

In a remarkable effort, the American Bureau of Ships designed a barge which, from its construction, became known as the 'Egg Crate'. The design took just a few weeks, and the first (named EC1) arrived at Falmouth for trials on August 10 1944. Fifty were ordered to be built in the United States, to be delivered at the rate of 14 per month, and a further 14 were ordered in the United Kingdom.

The Admiralty had learned some lessons from the trials of the 'Stirling Craft' and, since that time, eight 'Bangor' Class fleet minesweepers (180 feet long, but with twin screws and shallow draught) were converting to fleet tugs, specially to tow the displacement sweeps. Three of them—*Seaham, Eastbourne*, and *Fort York*, were ready for trials with the 'Egg Crates', as well as some of the big rescue tugs.

Trials were made with these 'Bangor' Class ships, and then with the tugs, using them first one, then two, then three in line abreast, and with differing combinations of the ship classes, showed that two large rescue tugs gave the best towing speed through the water (6.1 knots under full power). However, their deep draught again put these big tugs at grave risk, and so they were ruled out. Two of the 'Bangor' Class together gave 6.27 knots under full power, with an acceptable pressure signature, and with a third 'Bangor' standing by in case of breakdowns, which were expected to be frequent. The towing force required in the 'Bangors' to achieve this speed through the water was 35 tons.

The table below shows how narrow was the swept path achieved by the 'Egg Crates'. They could only be used as a proving sweep, not for clearance, and the path they swept was too narrow to be marked accurately by dan-buoys. The 'Egg Crates' took quite a bit of handling, too—two rescue tugs, in addition to the third 'Bangor', were needed in attendance when the barge was rounding bends in the channel, and acted as a brake. A swell-recording vessel accompanied the little convoy and a single L magnetic sweep cable, energised from one of the 'Bangor' tugs, was draped fore and aft over the barge to protect it against magnetic mines, while a pipe noisemaker towed 120 yards astern warded off the acoustic mines.

The 'Egg Crates' were bottomless, of cellular construction, with sparson bow and square stern, buoyancy being provided by the longitudinal tanks forming the sides. They had a length of 331 feet, a breadth of 64 feet, and a draught of 21 feet, while their displacement was 3,710 tons. No drawings or photographs of these craft have been found on either side of the Atlantic. They were not used operationally in the English Channel but later, early in 1945, in the swept channel between the North Foreland and Flushing they may have swept one mine. Designated the Fiftieth Minesweeping Flotilla, one of its units, EC7, was wrecked in a gale on the Belgian coast, but another, EC10, operated with some success from Ostend—at least it was towed safely to and from its sweeping area!

Seven 'Egg Crates' were sent from the United States to Japan for clearance operations late in 1945, but they do not appear to have had more success there than in Europe.

Concrete barges

These were the last displacement sweeps constructed in 1944–45. Designed to be easier and faster to tow than the 'Egg Crates', they were of two types—a light one, which would break into small pieces on being mined, but this one proved to be so flimsy that a bump on coming alongside a pier could sink it; and a stronger one, but this one turned out to be too heavy to float!

Two compromise barges (CB1 and CB2) were, however, constructed in the United Kingdom, and used for trials in the late summer of 1944. They were 190 feet long, 40 feet in breadth, and 18 feet in draught; they displaced 3,000 tons, and it was planned to build 150 of them. To be constructed in batches of 30, in the same building berths where the caissons for the Mulberry Harbour for the Normandy landings had been built, 25,000 men would have been needed, had the programme been carried through.

The optimum towing configuration with the two trial barges was found to be one 'Bangor' tug making 8½ knots through the water, but the width of the swept path was disappointing. All work on these barges was suspended in November 1945.

Depth of water	Swept path (in yards)	
(in fathoms)	'Egg Crate'	Concrete barge
6	60	60
7	50	62
8	40	53
9	35	59
10	27	0
11	20	0

Guinea pigs

These were merchantmen or small warships, filled with buoyancy material, fitted with remote control gear for the main engines, and used for proving runs in the swept channels approaching major ports. Operational experience with these ships differed widely. The British converted a merchant ship, *Formigny*, in the summer of 1944, but after some bad experiences, abandoned her in favour of the 'Egg Crates'. The Americans in the Pacific were more successful; damaged attack transports and 'landing ships tank' were used. Some Japanese ships were also pressed into this role in 1946, one being sunk by a pressure mine. The Germans used this type of ship for their main sweeping effort against pressure mines. They converted 117 merchant ships to the role of 'Sperrbrechers'. Their losses were heavy, both by mines and by air attack.

Despite all these valiant efforts, the official Royal Navy view in November 1945 was that speed restrictions were the only effective answer to pressure mines.

The German pressure mines laid in 1944 and 1945 were not very numerous and had little effect on the course of the war; but a much larger 'oyster' campaign would have posed a serious threat to the Allies. As it was, mines were the largest cause of shipping losses in the last year of the European war, both at the Normandy landings and on the Antwerp convoy route.

The RAF dropped many pressure mines around Germany in the last year of the war, causing numerous shipping casualties. German coastal traffic was becoming light, but some successes were scored against the new fast U-boats, types XXI and XXIII, which were working up in the Baltic. Many sweepers were also sunk.

The United States Air Force in the Pacific laid a very large number of mines around Japan in the closing stages of the war. They used a mixture of pressure and acoustic types. The Japanese minesweeping service was overwhelmed and over 200 ships, of more than half a million tons, were sunk in this area in the course of one year.

Assault sweeping

This was a new technique, developed by both the British and American navies for use in the many landing operations in Europe and the Pacific. There were four phases to an assault sweep: the passage to the landing beaches, clearing the approach for the assault, broadening out the beach-head, and then maintenance sweeping of the beach-head area.

Passage to the landing beaches

Here, there was a marked difference between European assaults, and those in the Pacific. In Europe, there would usually be only a relatively short passage from the fleet bases to the landing beaches, and Normandy may be taken as typical. The distance from the assembly area south of the Isle of Wight to the beaches was about 75 miles; and all of that had to be swept, ahead of the landing craft and support ships. Six parallel lanes were swept during the night before the assault; they were first marked with dan-buoys, carrying special lanterns, and later replaced by the much bigger steel channel buoys.

In the Pacific, the distances from the forward fleet bases to the landing areas were usually very much greater, always measured in hundreds of miles and often in thousands. In that situation, the sweepers as always needed to make their passage ahead of the fleet, in spite of any heavy weather that might be around. Sometimes there was still a long approach channel to be swept on arrival; at others they went straight into the clearance sweep for the assault itself.

Clearing the approach for the assault

On arrival in the assault area the sweepers would begin a fast clearance of all water through which the landing craft, including the bigger LSTs, would move during their run in to the beaches. This would often be done under heavy shellfire from the shore. While the battleships, cruisers, and destroyers could support the sweepers with their heavier guns from further offshore, a first objective of the sweepers would be to clear the anchorages and firing areas for the destroyers so that they could dart in and engage the shore batteries at visual range.

The big sweepers usually went in first, with wire sweeps out to cut any moored mines present. In Europe, this would mean the fleet sweepers in flotillas of eight ships each with two or more 112-foot motor launches leading the way ahead of the leading fleet sweeper, and with rope Oropesa sweeps out, making a skim sweep. Each flotilla would also have two or more danlayers attached to it, marking the channel as it was swept. Even at Normandy, not enough of the special Royal Navy 'Isles' Class danlayers were in operation, so that a number of fleet sweepers acted as danlayers for the assault, since many of the older danlaying trawlers were too slow to keep up with the sweeping formation.

In the Pacific, the big, fast destroyer sweepers usually went in first—in early assaults, the faithful old 'four-stackers', but later the new and fully-equipped 'Livermore' Class fleet destroyers. Then the influence sweeps against ground mines would follow; in both theatres of war, these were usually operated in the assault itself by the coastal sweepers, the excellent YMSs and MMSs. They would have their LL and SA sweeps operating and would sweep right inshore, to the shallowest depth of water in which they could operate. Although often they would be under fierce enemy fire and would have their sweeps cut by enemy shells, they must as often have parted their sweeps on inshore reefs and rocks, and even touched bottom themselves!

Destroyers and motor torpedo boats would move in with them as they swept, engaging the enemy batteries, since it was essential that the sweepers were able to keep exact station on one another, to ensure accurate sweeping of the area. Radio counter-measures would also be operated in the sweepers, to confuse the enemy shore radar in the hours of darkness before the assault.

Then came the landing craft, ploughing past the sweepers towards the beaches. Small LCVPs were often used as inshore sweepers, with rope sweeps for the very shallow water. Several good tries were also made at fitting the landing craft themselves as sweepers, to clear the last areas of water before they hit the beaches.

The Royal Navy fitted a number of 'landing craft, tank', Marks 4 and 5, with light Oropesa sweeps. They consisted of two marker floats from the old destroyer high-speed sweeps, fitted with diverter planes to carry the ends of the ropes out and away from the landing craft, as did the Oropesa floats and otters in the larger sweepers. These sweeps were towed on each quarter of the LCTs, using 65 fathoms of 1¾-inch sisal rope, and they could be handled by untrained personnel. They swept a path 75 yards wide, moving at 7 knots through the water.

Another idea was for clearing anti-tank minefields awaiting the assault forces in the shallowest water. In early 1944, the Royal Navy fitted an LCA with a rocket-propelled Bangalore torpedo—a tube, fired with a flexible hose attached to it, through which, after it landed, would be pumped liquid explosive from a one-ton container in the LCA. On trials, the rocket threw the tube 330 yards and it was filled with explosive in 25 seconds. On exploding, it made a crater in the sand eight feet wide and four feet deep. But it was a very risky operation and it was thought to be too dangerous to put the liquid explosive in an LCT, with the valuable assault tanks on board. General Montgomery did not like it and preferred specially-equipped army tanks to flail the sand after landing. So this remarkable idea, called 'Book-Rest' by the Navy, was abandoned.

However, the smaller LCVPs and LCMs were used successfully in both theatres of war. In the Pacific they were often carried, four to a ship, on the heavy davits of the APDs, the fast destroyer transports. So the landing craft roared in to the beaches, often narrowly missing the long sweeps, and many a time taking a wrong turn! Marker boats, PCs, PCEs, SCs, and MLs, flying large flags, guided them in through the swept water.

Broadening out the beach-head

Next, carefullly-planned sweeps would be carried out, some to clear manoeuvring areas for the bombarding ships, others to clear anchorages for the large attack transports and supply ships. This would be a hectic time, certainly lasting all the first day of the assault. Enemy shellfire would probably still be falling among the sweepers, and smoke to screen them would be laid by attendant small craft and by aircraft. The task of the sweepers would become increasingly difficult as the area became full of ships. It became more and more difficult to lay dans and keep their long LL tails clear of the anchored fleet. Single L tails, towed by a single sweeper rather than in pairs, often had to be used due to the traffic jams. By now, the sweepers would have run into any new nasty surprises laid out for them by the enemy. One in regular use by the Germans in Europe was the mine fitted with snaglines. The first type had up to 15 fathoms of thin cod line supported just below the surface of the water by cork floats, all painted green, to be invisible to the approaching craft. When a ship—or a sweep—touched one of the cod lines, it would activate an electrical switch in the mine and explode it.

The LCT rope sweeps were quite effective at first against these mines, so the Germans produced a superior type of snagline—it had a synthetic rubber core and was buoyant, thus not needing the easily spotted floats. Then another, and better type appeared, also during the Normandy invasion. This was rather like a certain kind of child's doll: a concrete sinker lay on the seabed, with an iron tripod protruding upwards from it, with a chemical horn on the top. The sinker had a 165 lb charge in it and when a sweep passed over the tripod the mine just rolled over, like the doll, and then returned to its dangerous upright position. It was called 'Katie'. A trial sweep against this mine (though it was abandoned) consisted of an old torpedo towing two diverters on the end of ropes, thus forming a double Oropesa sweep—though unmanned. It would run for a mile, at an average speed of 6 knots, but was too complicated to launch.

When the broadening out of the swept area had been completed during the day, the sweepers would spend a sleepless night. The big fleet sweepers would join the outer defence screen around the assault anchorage, ready to repel midget submarines, suicide boats, and swimmers. The smaller YMSs and MMSs would anchor in among the anchored ships to watch for mines falling by parachute from raiding aircraft and to plot their positions carefully, ready for sweeping the next day.

This was hazardous work, as the accounts of the Normandy and Okinawa landings show. At Normandy, for instance, no less than four of the 'Catherine' Class fleet sweepers (the American 'Raven' Class AMs) were lost on the night defence lines in just a few days.

Maintenance sweeping

The main task was to ensure that the whole beach-head area was clear of mines. Easier said than done. The problems of sweeping in a crowded anchorage became no less severe during this phase. Added to them was the hazard of the sweepers operating almost continuously—sweeping by day, and in the defence lines by night—and so becoming more and more tired, while losses of sweep gear of every kind often became very serious. Wire and LL sweeps parted from contacts with rocks or wrecks, or from non-comprehending friendly ships, faults in LL generators, and in main engines, hammer boxes going wrong—all could add up to the number of operative sweepers in any one area falling to an unacceptable level. This led to the requirement for an efficient maintenance and supply organisation, to keep the front-line sweepers going. The British and Americans took rather different steps to solve the problem. The sweepers needed a carefully-organised rota system of rest periods and maintenance periods, or the critical sweeping of the assault area would not be effectively carried out. Bad weather, too, had to be taken into account; the best illustrations of this were the great gale at Normandy, and the typhoons which hit the sweeping forces around Okinawa in the second half of 1945.

Maintenance sweeping, in its own way, also took in the special port clearance operations which were launched as the landings ashore took effect, and ports captured but infested with mines of all types needed to be cleared as quickly as possible.

Coming back to the crowded anchorages during assaults, the Royal Navy found it necessary to brief the big ships, the battleships, cruisers, and attack transports, on the problems which the sweepers faced in ensuring that the waters were clear of mines. Tables were issued showing the width of path swept in differing states of the tide, the time it took to shorten in a sweep, wire or LL, before a big ship could pass safely astern, and the sheer length of the wire and magnetic cables streamed by the sweepers, in the hope that this would help the big ships to avoid running the sweeps down.

Truly, assault sweeping was an art in itself, and one in which the front-line sweepers of both navies excelled—and let us not forget that the Canadian 'Bangor' fleet sweepers participated heavily in the Normandy landings and in the clearance of the French ports afterwards.

Headquarters and maintenance ships

From the early days of the European war, the Royal Navy had seen the need for specialised ships of this kind and, as the Pacific war developed, the US Navy followed suit.

Europe

As the numbers of minesweepers of any kind built up at the main minesweeping ports around the British Isles, the need was seen to have a Captain M/S in each major port, stationed ashore, to coordinate the operations and the maintenance. In the early years, while requisitioned trawlers in large numbers made up the sweeping force, old converted steamers were used as the headquarters ships, accommodating the Senior Officer and his staff. They were usually stationed at one base and, in the largest ports, the Captain M/S had also a 112-foot 'B' Class ML at his disposal. The fleet sweepers in early days were usually based at naval ports and, with their Senior Officer embarked in one ship, they relied on the dockyard for repairs. For example, at the Nore was the old Isle of Man steamer *St Tudno,* berthed in the Sheerness port area. Of 2,326 tons gross, and built in 1926, she did sterling service there. At Scapa Flow, similarly, was the *Ambitious,* built in 1913 as the *Algoma,* and of 1,849 tons gross. A little later in the war, when the mine destructor ships were losing favour, they were seen to be suitable for this role. *Borde* was converted and went to the Mediterranean, where she provided valuable experience in the needs of such a ship, operating far away from a main fleet base.

Normandy invasion, 1944

This produced the first really major need for supply and repair ships for a large force of sweepers operating for weeks on end under pressure and away from their home base, even though the latter was only some 100 miles away.

The Royal Navy brought *Ambitious* down from Scapa and sent her over to the British assault area, while the US Navy sent over a coastal mine planter, *Chimo*, for the same purpose, and used ships of the same category in the Pacific. The Royal Navy also had the converted *Kelantan* anchored at Spithead, to act as an M/S storeship, with three old drifters ferrying spare sweeping gear over to the beach-head.

When the assault on the island of Walcheren, at the mouth of the Scheldt River leading to Antwerp, was carried out in November 1944, the old *St Tudno* came across from Sheerness to lead the flotillas of motor minesweepers, and stayed with them during the clearance of the river itself, and after the war, during the mine clearance operations up the Dutch coast.

Far East, 1944–45

During this period, the Royal Navy was planning its Fleet Train to serve the two big fleets in the Far East, and learning from the great experience of the US Navy in this field. Clearly, support ships for the sweepers were essential, but the Senior Officer needed to be up with his sweepers, so the role was split.

M/S headquarters ships

For the vast distances in the Indian and Pacific Oceans, two mobile headquarters ships were required, one for each Fleet. So two of the later units of the 'Algerine' Class to be completed, were specially fitted out for this role. *Niger*, for the British Pacific Fleet, arrived out on station before VJ-Day, but *Fierce*, for the East Indies Fleet, only arrived after hostilities had ended. Both were badly needed for the great postwar mine clearance. They received extra cabin space aft, in place of the LL reel, though the Oropesa winch and sweep gear was retained. The SA sweep was also removed, and these ships probably did little sweeping themselves.

Two extra long motor launches were carried on special davits, and extra radio-telephone lines were installed. The close-range armament consisted of four single 40 mm Bofors, following the Pacific gunnery policy, but otherwise they remained standard units of their class. It is interesting that, even at this late date, these two ships did not receive the new 4-inch Mark XXI gun, specified for this class.

M/S store and repair ships

The battle plan for the two fleets showed six fleet sweeping flotillas of 'Algerines' for the East Indies Fleet, and the British Pacific Fleet was to have two flotillas of this class, plus one flotilla of the Australian-built 'Bathurst' Class, and two BYMS flotillas. For these sweepers, four or five store and repair ships were designated. Each would be able to service two flotillas, and four of the old mine destructor ships were got ready, *Corbrae, Andelle, Borde,* and *Fairfax*, together with the old *Kelantan*. But the end of the war happily came soon, and only *Corbrae* and *Kelantan* went out.

For this service, their LL generators and magnets were removed, a stern anchor fitted for use in isolated anchorages, and their range of operation increased to 3,500 miles. Derrick-lifting capacity was added for handling the heaviest generators fitted in the sweepers, and the maximum number of dan-buoys and their gear was to be carried. Spare LL tails were shipped on special reels, and many spare Oropesa floats and wires stowed below. Other refinements included a foundry, a recreation room for the sweepers' crews, and a very large sickbay.

The US Navy was using similar ships throughout the Pacific, to service their very large numbers of fleet and yard minesweepers. Wonderful work was done in this way in keeping large numbers of sweepers in service despite the worst that continual sweeping and heavy weather could do to the sweep gear—or to the sweeping crews themselves. A new cruiser-minelayer, *Terror,* was converted as a minesweeper flagship, and gave splendid support in the Pacific.

'Algerine' Class fleet minesweepers (RN)

This was the last class of fleet minesweeper built for the Royal Navy. To some extent, these ships were still a compromise, but they, with the American 'Raven Class, represented the best in minesweeping thinking to 1945. They were fine ships and gave excellent service during the later stages of the war and during the postwar mine clearance operations.

Design development

The 'Halcyon'/'Hebe' Class, built before 1939, produced the first specialised fleet minesweeper since 1918, and at the time when the 'Black Swan' Class sloops took off on a different route, covering anti-submarine and anti-aircraft requirements. The 'Algerine' Class was a logical development from the 'Halcyons', incorporating many of the lessons of the early war years, and rectifying the lack of space in the intervening 'Bangors'. This new class still retained much of the prewar roominess, with partly planked decks, and the 16-foot sailing dinghy retained in its own special davits—a surprising luxury in wartime. They were originally designed as wire sweepers only, but they finished up operating all types of sweep, and it is a credit to the original design that they were able to do this without any overcrowding. Indeed, it could be argued that their sweep deck

Coquette, *of the British 'Algerine' Class, with her Oropesa sweep ready for streaming over her stern. The big steam sweep winch can be seen on her sweep deck.*

was the best laid out and the roomiest of any of the classes of fleet sweeper, even after the LL reel and the explosive sweep had been added. On top of this, they retained their full depth-charge rails, and four throwers. This was a considerable achievement.

The Director of Minesweeping included in the 1940 and 1941 programmes an order for 'Repeat Hebes'. They were to be similar to the 'Hebes' (the later ships of the 'Halcyon' Class), but with a simplified armament and an extra two feet on the beam. It had been found in wartime that the earlier class had difficulty in carrying an increased load of fuel, stores, and sweeping gear, and this new class overcame most of those problems. There was some argument at the time as to whether this new class was too large and luxurious for mass-production in wartime. Lobnitz, one of the main builders of fleet sweepers, and located on the Clyde, designed another class, midway in size between the 'Bangors' and the 'Repeat Hebes', but the latter had been progressed rapidly,and this design was chosen for urgency of construction.

The 'Algerine' Class name was added while the early units of the class were on the stocks and, of all the Royal Navy classes, the names chosen for this class were perhaps the most colourful, with traditional Royal Navy names from the past coming through strongly. It was a cheerful sight to see a flotilla of this class passing by, with such a collection of names as *Brave, Fancy, Plucky, Rosario, Fly, Rattlesnake, Stormcloud,* and *Welcome.*

Production

Orders for this class were first placed on May 16 1941, and thereafter flowed thick and fast. They were built under *Lloyd's Register* survey, and were economical units—even in April 1943 the quoted price averaged £171,000, made up of £101,000 for the hull and £70,000 for the machinery.

A total of 97 units of this class was completed, with 21 ships cancelled in addition. The 1940 programme contained 20 ships and, a sign of the wartime shortages, only three of these were to have reciprocating engines, the other 17 having turbines; whereas of the 26 ships in the 1941 programme, only seven had turbines.

Of the total of 97 ships, as many as 49 were built in Canada as part of that country's truly remarkable shipbuilding effort. Fourteen of these were ships built under original orders from the US Navy, and turned over to the Royal Navy under the Lend-Lease programme. The Royal Canadian Navy retained only 12 units of this class for itself, and fitted these out as anti-submarine escorts.

The 1942 programme included 14 ships, but the 1943 programme reflected the Canadian contribution; for a total of 36 'Algerines' were ordered from Canada in that year—19 direct orders for the Royal Navy, and a further 17, in exchange for 15 UK-built 'Castle' Class corvettes, and three 'Loch' Class frigates. Ten of the units cancelled from orders placed with Harland & Wolff in Belfast were linked to this; for by this switch of orders the maximum number of the new frigates could be built in the UK yards, while the Canadian yards concentrated on the building of the 'Algerines' together with corvettes. By this change, not only was the building of the frigates accelerated but the Royal Navy received two flotillas of the valuable 'Algerines' a full year earlier than it otherwise would have done. This was a fine example of Anglo-Canadian cooperation.

Apart from the 12 units retained by the Canadians, all units of this class were commissioned and operated by the Royal Navy, and none was transferred to other navies. Nine ships of the class were lost by enemy action, five being sunk and four written off as constructive total losses.

These ships retained the strong naval appearance started by the prewar sloops. The bridge was a full naval design, and a full Asdic outfit was carried. The funnel became larger, with an 'Admiralty top', and the mast was a tripod, with two yards, to carry the many minesweeping flag signals. They were fitted with radar from the outset—Type 271 at the back of the bridge, and Type 290/291 at the masthead. In refits from 1944 onwards, the Type 271 was removed, and replaced by the searchlight, which had been on a high platform above the radar; the smaller aerial of the Type 268 was mounted on a spur at the lower yard level.

Armament

This became quite strong; a single 4-inch Mark V on an HA/LA mounting, was put on the forward deck. This was an old gun, fitted with a new shield—the new 'Loch' Class frigates were similarly armed; but it was never more than a temporary allocation. The 'Algerine' Class was to receive the new 4-inch Mark XXI gun, but this came out too late and was never fitted. Only one gun of this mark ever went to sea, and this was in the last ship of the 'Loch' Class to be completed, *Loch Veyatie.*

The close-range armament, on the bridge sponsons and on the gun deck aft, mounted over the LL reel, was usually of Oerlikons. In the early days the single 20 mm mounting was fitted, but later ships received the excellent hydraulic twin mounting, which included a shield for the gunlayer. The 'Algerines' seemed to do pretty well in the allocation of these popular mountings, compared with the North Atlantic escorts! In 1944–45, a number of ships designated for the East Indies and British Pacific Fleets had their Oerlikon mountings removed and replaced with single 40 mm Bofors.

For mine clearance two anti-tank rifles of the Boyes type were also supplied, for sinking floating mines cut by the wire sweeps. These were mounted on portable brackets on the bridge Oerlikon platforms. But as the guns were unshipped before entering harbour, they are not usually visible in photographs. They made an ear-shattering noise, but their armour-piercing bullets were much more effective for this work than the more usual 0.303-inch machine-guns.

Boats

The boat outfit was surprisingly lavish, a naval 25-foot motor boat on the starboard side, and a 27-foot whaler on the port side, both in gravity davits. We may compare this with the trawler type dinghy, which was all the later corvettes got until near the war's end.

Sweeping equipment

The squared-off stern was ideal for handling the wire sweeps. The big LL fairleads were in the centre, flanked by those for the Oropesa sweeps, with the davits mounted right on the edge of the quarters.

A technique was devised for the streaming and recovery of a combined Oropesa and LL sweep in this class, but it was not often used in operations. The big minesweeping winch took pride of place at the forward end of the sweep deck, with the LL reel stowed just forward of it under the Oerlikon gun deck. The generators for the LL sweep were in a room on the deck below, and the exhausts for these machines were taken past the LL reel to the break of the main deck. In photographs taken of this class from 1944 onwards, a black stain can often be seen on the hull at this point, where the exhaust fumes came out. The LL reel weighed 30 tons when loaded with its cable and, while the 'Algerine' Class could accept this extra weight, with a fore and aft length of 17 feet on the reel itself, the earlier 'Bangor' Class could not, and were thus poor influence sweepers.

The SA acoustic sweep was of the towed box type, and was handled by a small derrick mounted under the bridge on the starboard side. The box was stowed on the deck beside it. The explosive sweep protective shield was mounted on the guard rails right aft, usually on the starboard side of the sweep deck, forward of the Oropesa floats.

The stock of dan-buoys, with their colourful flags, was stowed horizontally on a rack over the minesweeping winch, with the reserve supply carried on the uncluttered deck by the funnel.

One further sign of the follow-on from the 'Hebe' Class should be mentioned—all 'Algerines' carried two teak accommodation ladders, complete with davits, just aft of the break in the main deck. They looked very smart, when lying at anchor as a flotilla, between sweeping operations.

Speed of building

There were two main builders of the class—Lobnitz, on the Clyde, for the reciprocating engined ships, and Harland & Wolff, in Belfast, for the turbine-engined. In Canada, there were also two main builders of the class—Redfern and Port Arthur.

In the United Kingdom, Harland & Wolff, who built as many as 24 units of the class, achieved a fastest time of seven months and three days—as good as the

315

118

fastest Canadian building times. Lobnitz, with 18 ships, had a fastest time of eight months and 17 days. To show how building times fluctuated greatly, the slower times for these two yards were of the order of 15 to 16 months.

In Canada, building times for this class also varied quite widely, but were comparable with those of the UK-built ships. These ships were largely of riveted construction, and it is interesting to compare the faster times achieved in the United States, for the 'Raven' and 'Admirable' Classes, where welding was used so extensively.

'Raven' Class fleet minesweepers (AM—USN)

This class represented the latest in US Navy minesweeping thinking to 1945 though unlike the Royal Navy, the Americans went on after the war to produce other designs of fleet sweeper, albeit under other class designations.

The 'Raven' Class, with the British 'Algerines', was undoubtedly the outstanding fleet sweeper design of the war. It is noteworthy that through the Lend-Lease programme, 22 units of this class were transferred on completion to the Royal Navy, where the enthusiasm was fully shared. This class was prominent for its speed, handiness in manoeuvring, and standard of equipment and accommodation. The Royal Navy retained its preference for the fast-running, steam-powered minesweeping winch, whereas the US Navy went for electric power throughout, but otherwise, thinking was remarkably similar.

Design development

The first two units of this class, *Raven* and *Osprey,* came out in 1940. They represented in part the reawakening of US Navy interest in minesweeping, prompted by Royal Navy war experience. This was a full year before Pearl Harbor. It may be noted that the basic design did not change, although this class was in production right through to the end of 1945; so the comparison is with the 'Bangor' Class, as well as with the 'Algerines', and this demonstrates how fast and how effectively US Navy design moved, once the impetus was given in 1940.

There had been no new US Navy fleet sweeper design since the 'Bird' Class of World War 1, so that there were no inhibitions to moving towards an entirely new design, and that is what the Americans did. The result was quite distinctive among all the escort and sweeper designs of World War 2.

The hull had a turtle-back fo'c'sle deck—the only one in this period, though it reappeared after the war, not least in the Royal Navy's first postwar frigate designs. The fo'c'sle head had a steel bulwark round the stem, which also was unique in warships of this period. Next, the fo'c'sle deck stopped at the after end of the bridge in UK (but not US) destroyer fashion, so that there was a break to the fo'c'sle, and a side deck running aft on each side of the superstructure, right to the forward end of the sweep deck. This meant that there needed to be sponson decks built out over the side deck, to carry the boats and the short-range weapons—a marked difference from the 'Algerine' Class. The bridge rose in two decks from the fo'c'sle deck, and was remarkable in that a single Oerlikon was

Above left Chief, *of the American 'Raven' Class, showing her sweep deck lay-out, with the SA derrick amidships.*

Left Steady, *of the USN 'Raven' Class, off the Sicilian landing beaches in July 1943. She has no normal funnels, and the SA gear is mounted on the stern.*

mounted on the upper bridge. The mast was a pole, with a single high yard, and the excellent American radar, Type SL, mounted right at the tip of it. Two searchlights were mounted on a high platform over the upper bridge; it is worth noting that these searchlights were seen to be needed in this position both in this class, and in the British 'Algerine' Class. In escorts, the American DE classes had them on the wheelhouse level, and 'River' Class frigates one deck higher, but neither had them mounted as high as these two classes of fleet sweepers.

The motor whaleboat, of the same length as the Royal Navy's motor boat, was housed on a sponson on the starboard side, and handled by a boom (derrick), and not by davits; this practice was followed in the later 'Admirables'.

There were two prominent funnels to this class. Again, this was a unique feature, except in the turbo-electric DE classes, and there the twin funnels were trunked together. A short pole was usually added to the after funnel, to carry the radio aerials.

Armament

Here, the excellent US Navy equipment was available for this class, with a marked difference to the RN classes of the day. The first two ships, completed at the end of 1940, carried two of the new 3-inch 50-calibre guns, one forward of the bridge and one aft above the sweep deck. Once the class went into quantity production, the after gun was soon replaced by single, and later twin, 40 mm Bofors, and single 20 mm Oerlikons were mounted on the bridge and amidships. Further single Oerlikons were mounted on the sweep deck (fantail) in some ships.

The anti-submarine equipment was impressive and must have been much in demand when these ships were pressed into service as long-range escorts in the Pacific. Unlike the 'Algerine' Class, the 'Ravens' were able to carry a full Hedgehog ahead-throwing weapon, forward of the bridge, but just aft of the forward 3-inch gun, following the excellent layout achieved in the destroyer escorts. Aft, the early ships of the class only carried depth charges in single traps, as in the RN's 'Bangor' Class. This was soon revised so that full depth-charge rails, together with throwers, were carried. The sweep deck was not as spacious as that in the 'Algerine' Class and the layout was, therefore, more crowded. The basic layout of the sweep deck was very similar to that in the British ships, with the big minesweeping winch (in this case, electrically driven) taking pride of place, with the big LL reel forward of it, and with the minesweeping davits on the quarters. The acoustic sweep was, in some of the earlier ships, mounted on a boom over the bow; but this was soon changed to the towed box version.

Production

Production of this class was split effectively into three parts. First, the two experimental ships appeared at the end of 1940. Then, 51 ships were constructed, launched between the beginning of 1942 and the end of 1943; then a further 20 ships were launched between late 1944 and early 1945. A further 22 ships were built for the Royal Navy, under Lend-Lease.

So the total units of this class built numbered 95, of which 73 were commissioned by the US Navy; a further ten units were cancelled, and there was some complicated allocation and reallocation of ships between the USN and the RN during the construction period. Thirteen units allocated to the RN were finally retained by the USN, as the minesweeping requirements in the Pacific grew, and these were in part made good by the reallocation to the RN of 'Algerine' Class units which had been ordered by the Americans from Canadian shipbuilding yards.

Seven ships of the class in US service were lost. The class gave excellent war service all over the world, since in addition to the Pacific operations, units of this class were presented at the Normandy landings (where two were lost), and in the Mediterranean during the assault operations from Sicily up the Italian coast, and at the assault on the south of France.

Eleven shipyards in the United States participated in the building of these ships, and some very fast building times were achieved. Four months, 21 days seems to be the record, marked up by the General Engineering and Drydock Corporation, of Alameda, California, in building *Pheasant* (AM 61). The mass-production methods brought in for the construction of the destroyer escorts rubbed off on this class too; but units completed later in the war had much longer construction times, as the pressure eased off.

'Admirable' Class fleet minesweepers (AM—USN)

This was the second of the fleet minesweeper classes constructed for the US Navy during the war. The class filled part of the pressing need for more, and ever more, fleet sweepers; but they were not seen as effective sweepers in the same category as the 'Ravens'. They were smaller—184 feet overall, against 220 feet in the earlier class; and their speed was 14.8 knots running free, as against 18 knots.

Design development

Part of the fantastic American shipbuilding effort of the middle war years, the design for this class was rushed through. The lead design yard turned out to be the Pullman Car Company, as unlikely as you could imagine; but they had excellent design, planning and procurement skills and experience.

The same hull design was used for an escort version, the patrol craft escort and, in fact, it was this version which was built by the Pullman Car Company itself, in its shipyard near Chicago.

Some criticism was later levelled at the design, but this takes nothing away from what was by any standards a remarkable shipbuilding effort.

Production

A total of 124 ships of the class were completed in the minesweeping version, or intended for it. Of these, 34 were transferred to Russia, after some service in the US Navy, and four others were similarly transferred to Nationalist China. A further 51 ships were cancelled, seven of them after the hulls had been launched.

The escort version had a similar history; and there, 150 were originally ordered by the Royal Navy, but only 15 were delivered in the end, due to the progress of the anti-submarine war.

Design

This was quite different from that of the 'Raven' Class. With a shorter hull, accommodation needs required the fo's'cle deck to be carried right aft to the forward end of the sweep deck—much the same layout as in the RN's 'River' Class frigates. A very large bridge structure was built over this, which made these ships look smaller than they really were; but this gave an excellent open bridge for controlling sweeping operations. A vertical pole mast was fitted, with supporting struts forward to the bridge structure. A distinctive feature was that two gaffs were fitted to this mast, one forward and one aft, making this class quite unique.

The excellent and ubiquitous Type SL radar(or one of its close equivalents) was fitted right at the top of the mast, and these ships received the same high-class radio-telephone equipment as the destroyer escorts—far ahead of their British counterparts.

The main engines were diesels with direct drive, and the engines coupled to the shafts through simple single-reduction gearing. This was a good arrangement, but the layout was somewhat marred by the main engine exhausts being at first led to the atmosphere through exhausts in the hull, low down amidships. This gave difficult ventilation problems, since the main intakes for the crew's living accommodation were at uppper deck level, just over these exhausts, so that, especially when the ships were lying together in harbour in nests, it was all too frequent that the diesel fumes were drawn into the accommodation. Later units had the diesel exhausts led to the upper deck and the fumes discharged through thin funnels; later again, small streamlined funnels were added to keep this under control.

Armament

As with all American war-built ships, this was of a very high order. The excellent 3-inch 50-calibre single gun was mounted on the forward deck, and the

Execute, *of the American 'Admirable' Class, in November 1944. Though smaller than the 'Raven' Class, these ships managed to have an efficient sweep deck, with a heavy gun armament.*

Hedgehog A/S weapon was mounted just aft of it—no British fleet sweeper managed to mount this weapon. Aft, two single 40 mm Bofors were mounted in tubs, on and forward of the sweep deck; later, these mounts were changed to two twin mountings. Single Oerlikons were mounted around the bridge; early units of the class had two mountings and this, too, was increased until four or five single mountings were included.

Sweep deck
This was well designed and relatively roomy. Full sweep gear was carried including the Oropesa sweep, the LL sweep, and the SA towed box sweep.

Boats
One boat was carried, the standard US Navy 26-foot whaleboat. This was stowed on the upper deck on the starboard side, and was handled by a large boom, mounted at the foot of the mast.

Seaworthiness
These ships served all over the world and kept up with their warship colleagues throughout the Pacific, weathering splendidly many a storm and even typhoon. It is, however, worth recording that their flat bottom and heavy topweight resulted in their being prone to rolling in a seaway, at times, to a truly remarkable degree. In those ships transferred to the Royal Navy, for instance, even old seadogs from commercial trawlers used to operating in winter off Iceland found that they were seasick as never before.

Production
These ships were completed from early 1943, right through to the end of the war. Launchings were 21 in the latter half of 1942, 60 in 1943, and 40 in 1944. The fastest building time recorded for the minesweeper version was six months, 17 days; but many were slowed down while on the stocks, when shipbuilding was run down in 1944–45.

Chapter 6

Assault on Europe, 1944–1945

The first four years of mine warfare in Europe were a constant game of technological advance and counter-measure between the British and German minelaying and minesweeping forces. But 1944 heralded the greatest laying and sweeping operations of the lot. No longer were the major efforts put into mining coastal channels to block ports and stop the flow of shipping; now the well-prepared assaults on Europe were launched with maximum effort. The secret pressure mines were laid by the Germans and there were intensive Allied minesweeping efforts using an enormous fleet of sweepers.

The build-up to Normandy

The first half of the year was by no means quiet in home waters; indeed, the mine totals for the year (without any of the Northern France happenings) were well up on the year before—and most of this took place before all attention was focused on the landing beaches. The Germans kept up the mining campaign, to try to catch the invasion fleet as it assembled in the ports on the South Coast of England. E-boats laid a field in the main approach channel to Portsmouth in January; a very difficult clearance operation bagged 14 ground mines, but the effort that went into this clearance was considerable and showed how far the sweepers had come since 1939. First, a flotilla of BYMSs carried out an exploratory Oropesa sweep, in 'G' formation; they were followed by a flotilla of MMSs, sweeping with LL/SA, in four sub-divisions of 'Q' formation. Behind them came MLs as danlayers, then the whole operation was repeated quite a few times, as it was thought that the mines might have an arming delay of as much as six days. The BYMSs also acted as LL/SA sweepers, and then the Ninth Flotilla of fleet sweepers joined in and completed the wire clearance sweep. All this was carried out in fog and with a force seven storm running. The mark boats even had to use searchlights to show their positions.

Then a month later, a field was laid outside Lowestoft. In the first BYMS clearance on their own, the 157th Flotilla of seven ships, with four old trawlers as danlayers, and MLs as skim sweepers, cleared the field in eight days. The BYMSs first swept with Oropesa and SA, then with LL/SA, and cut 23 'R' type mines. It was a long operation, since their passage out and back from their anchorage to the field took 5½ hours, and they lost many sweeps in the shallows off Smith's Knoll shoal.

German minelaying in the Channel

Then, at the end of May and just before the assault, E-boats in two groups laid a small field off Newhaven close to the busiest route for the invasion shipping. The number of mines in the field was small, but it still was troublesome as it extended over 55 square miles of sea. A mixture of sweepers descended on this field for a quick clearance before the assault started. The Eighteenth Flotilla of 'Algerines', and a number of BYMSs and both short and long MMSs arrived on the scene, which was 15 miles offshore, in the coastal shipping lane.

The 101st Flotilla of MMSs first swept in three divisions, in 'Q' formation, covering the ground ten times, first with LL, then with SA, and then with the two combined. The BYMS's came, sweeping in 'H' formation, with double Oropesa, and also using their towed acoustic boxes, SA Type A Mark I. Between them, they swept ten times, scattered widely in the area.

Last came the fleet sweepers, using influence sweeps as they were now beginning to do; they lifted three acoustic mines, one of which scored a near miss on *Stormcloud*.

Three MMSs from Portsmouth acted as danlayers for the sweeping force, but the storm which was blowing shifted most of their dans, while the spring tide which was running was so strong that the dans were often submerged, even with double pellets attached for buoyancy.

After the assault on Normandy started, German minelaying in British coastal waters fell to a low level, as their main efforts were concentrated against the beach-head. This was just as well, when it is seen how many sweepers had to be diverted to work with the assault forces. But the Germans were also suffering from a shortage of mines, due to the RAF bombing raids, so that apart from a defensive moored field in mid-Channel, they largely conserved their supplies to fight off the invasion.

The invasion of Normandy

The minesweeping side of this great operation has never before been told in all its exciting detail; this is surprising, since the role of the sweepers was critical to the success of the assault, and the greatest force of sweepers ever assembled was needed to do the job. In the event mines turned out to be by far the greatest naval threat to the invading forces, especially when the U-boats failed in their bid to make a mass attack on the fleet.

All the water to be used by the invasion convoys had first to be swept; as with every invasion, this put them away out ahead of all the other warships. It was inevitable due to the shortness of the midsummer night, that many of the sweepers would be in full view of the German defending forces for some hours before the assault commenced. Add to that the need to sweep ten new channels south from the Isle of Wight to the landing beaches, in the hours of darkness, but with very high standards of accuracy, and you have a major minesweeping problem!

The minesweeping forces

A thorough study of all the relevant records shows that 306 minesweepers were present at the assault; this figure varies from other accounts, but does seem the correct one. Of that number, only 32 were from the US Navy; this is not surprising as the naval forces for Normandy were, by agreement, predominantly British. The Americans were in the middle of their island-hopping campaign in the Pacific and most of the new warships, including many minesweepers, pouring out of American shipyards were badly needed there. But while only 32 of the

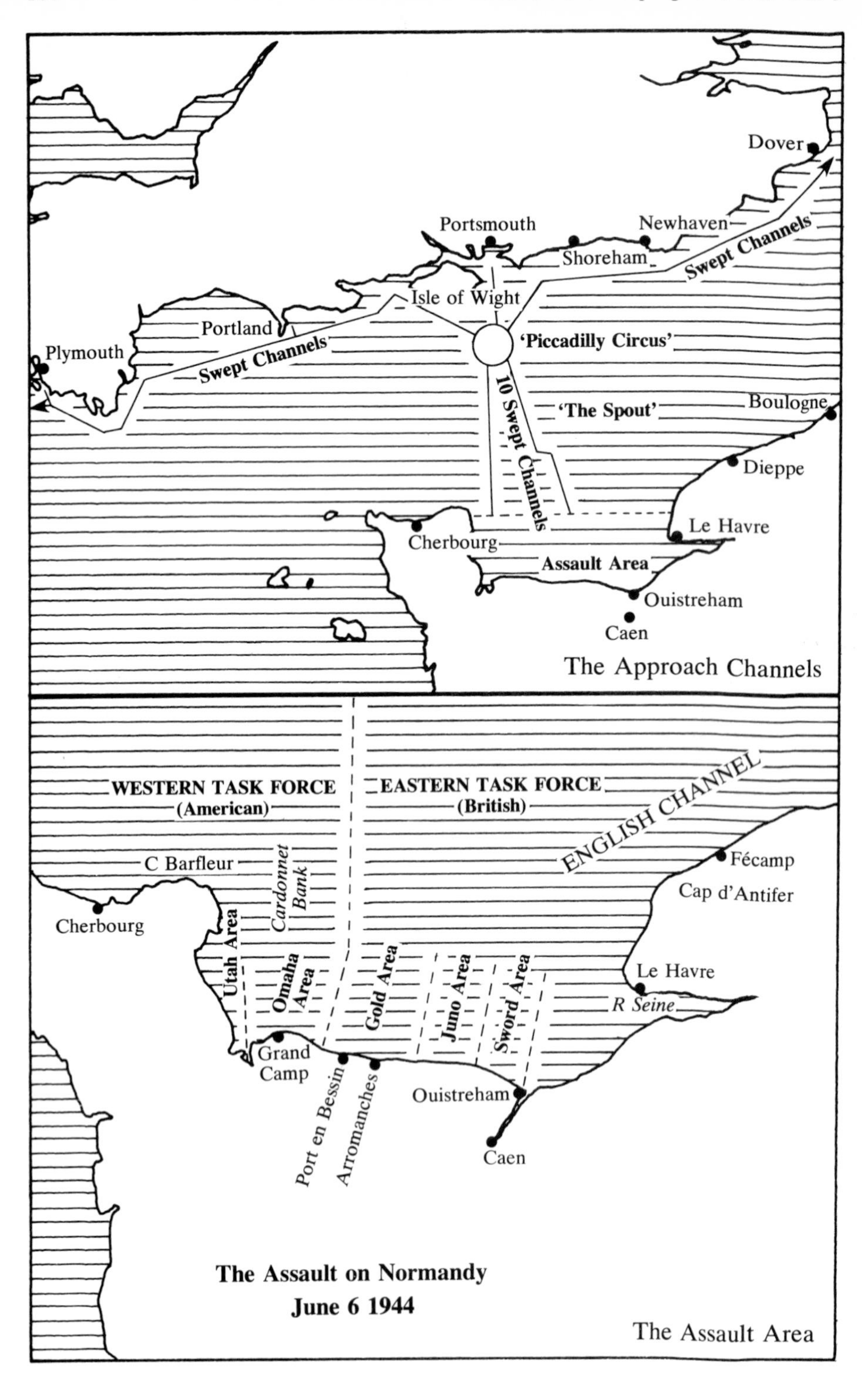
Dover
Portsmouth
Newhaven
Shoreham
Swept Channels
Isle of Wight
Portland
Plymouth
Swept Channels
'Piccadilly Circus'
10 Swept Channels
'The Spout'
Boulogne
Dieppe
Le Havre
Cherbourg
Assault Area
Ouistreham
Caen
The Approach Channels
WESTERN TASK FORCE
(American)
EASTERN TASK FORCE
(British)
ENGLISH CHANNEL
C Barfleur
Cardonnet Bank
Fécamp
Cap d'Antifer
Cherbourg
Utah Area
Omaha Area
Gold Area
Juno Area
Sword Area
Le Havre
R Seine
Grand Camp
Port en Bessin
Arromanches
Ouistreham
Caen
The Assault on Normandy
June 6 1944
The Assault Area

sweepers flew the US ensign, as many as 84 of them had been built in America, all the balance flying the Royal Navy's ensign under Lend-Lease.

The Royal Canadian Navy played a significant part in the invasion, as it was already doing in the North Atlantic; 15 of the 'Bangor'Class fleet minesweepers off the beaches were Canadian built and manned. But most impressive of all was the massive Royal Navy minesweeping force—274 ships of all types; and if we examine the breakdown, we see the great change which had taken place in the types of British minesweeper in the preceding four years. Of these 274 ships, only 43 had been in existence at the outbreak of war, and of those 25 were danlaying trawlers, some of great age.

In the fleet minesweepers, by far the largest single class was the 'Bangor', with 45 ships present, including the Canadians; these handy little ships proved their worth again at this time. But 25 of the new 'Algerine' Class were also there. It may seem surprising that they were not more numerous since, by VE-Day just over a year later, the Royal Navy had in commission about 100 ships of this excellent class. At the time of the invasion, this class was still being built in great numbers, and the 25 were all that were available. Two other flotillas of the same class were busy in the Mediterranean, and the remaining one was retained at Harwich and was thus the only fleet minesweeping flotilla in home waters not thrown into the assault.

One result of this, was that to make up the number of fleet minesweepers needed, two flotillas of prewar built ships were included—the First Flotilla of the 'Halcyon' Class, built just prewar, and only recently back from North Russia, and the prototype in many ways of the 'Algerines'; and the Fourth Flotilla of the 'Town' Class, built at the end of the World War 1, and nicknamed 'Smokey Joes', as they were, incredibly, still coal burners.

One should note the importance, too, of the coastal minesweepers, both British and American built. All of wooden construction, they underlined the campaign against the influence mines, and numerically, outnumbered even the fleet minesweepers. One can add to that yet another remarkable change in the balance of minesweepers since 1939; then, a vast array of commercial fishing trawlers was requisitioned for wire and influence sweeping but, while many of them were still serving in UK waters, not one of them was included in the invasion force in a sweeping role. But 25 of them were in use as danlayers, and a further half-dozen for transporting replacement sweeping gear to the flotillas on the far shore.

In passing, the effect on the UK-based minesweeping forces of the disappearance of so many sweepers for the invasion should be noted. All of the Royal Navy sweepers were drawn from flotillas based on British ports; there remained for routine sweeping of the vital swept channels around the British Isles just the following: eight 'Algerine' Class fleet minesweepers, 28 BYMSs, 166 MMSs, 51 war-built trawlers, 309 prewar trawlers and 57 prewar whalers and drifters, making a total of 619. One-third of the total minesweeping force, including nearly all of the fleet minesweepers, and one-third of the vital coastal minesweepers, had been withdrawn for the invasion.

The danlayers for the assault also show some interesting trends. The use of three or four danlaying trawlers for each fleet minesweeping flotilla was now an accepted part of mine warfare tactics, and this was fully implemented for the invasion. But at this time, very few of the specially constructed or converted vessels had been completed, so that only six of the Royal Navy's converted 'Isles'

Class were present at the assault, while a year later, about 40 of these were in commission or completing.

To make up the danlaying force eight war-built fleet minesweepers were used in this role at Normandy, plus the 25 prewar trawlers, referred to above, each carrying 70 dans. The coastal minesweepers, BYMSs, YMSs, and MMSs, did their own danlaying, two ships from each flotilla being so designated. All the danlayers attached to fleet minesweeping flotillas were British.

Another special feature of the assault minesweeping force was the use of three or four British 112-foot 'B' type motor launches, to carry out a skim sweep ahead of the Senior Officer's ship of each fleet minesweeping flotilla, this sweep being the only one, by tactical definition, operating in unswept water. A total of 36 Royal Navy MLs were used in this dangerous role, which had assumed importance in assault sweeping in both the Royal and US Navies in the previous year.

The US Navy used SCs (submarine chasers) in the same role on occasion, but though 18 of these craft were operating in the US assault sector, it is probable that they only worked as escorts and landing craft guides in the boat lanes, since of the 36 Royal Navy minesweeping launches, 16 were allocated to the American assault forces.

A last point which should be made here is that the totals of minesweepers shown in this section do not include any of the landing craft, fitted and used during the assault as snag line sweepers. Records do not show just how many landing craft so fitted were in the Normandy assault, though it can be seen that 16 of them were grouped as Sweep Unit 7 in the US sector. Certainly a good number must have been operating in this way, in both sectors.

Some special arrangements

Combating German shellfire

During the initial assault, many sweepers would be in sight and gun range of German batteries for considerable periods. A lot of work went into preparations to combat this, and much of the effort proved successful in the assault itself, when the sweepers had to maintain their formations accurately.

First, some of the bombarding cruisers were earmarked to dash up to sweeping flotillas which came under fire and engage the German batteries in counter-fire. Next, radio counter-measures were used to confuse the Germans' radar gun control. Two ships in each fleet sweeper flotilla received this equipment, as did a number of BYMSs and MMSs. Lastly, smoke was used effectively, and escorting coastal craft, or patrolling aircraft, were employed to shroud the sweepers when they came under fire.

Headquarters and supply ships

In spite of the short sea distances involved, both the British and American sweeping forces needed headquarters ships right up with them at the beach-head. These ships held big stocks of spare sweeping equipment and could carry out emergency repairs, to save the ships returning to the UK.

The Royal Navy had *Ambitious* stationed at the beach-head, and *Kelantan* at Portsmouth, the latter to co-ordinate the turnround and repair of returning sweepers. Spare equipment was ferried over to the far shore in old trawlers, rejoicing in names like *Star of Freedom,* and the Portsmouth organisation could handle simultaneously two fleet sweeper flotillas, one BYMS and one MMS flotilla. The US Navy brought over a converted mine-planter, *Chimo,* with three YMSs to help her in transportation of stores.

Danlaying

The accurate marking of swept channels was vital to the success of the great sweeping efforts; and the system well tested by the Royal Navy in UK coastal waters was developed further for the assault. Measurement of distances was done by Taut Wire Measuring Gear, fitted in a number of fleet sweepers and danlaying trawlers. It was based on paying out fine piano wire, and was very accurate. In addition, all sweepers were to keep a plot of the dans as they were laid, to pick up any inaccuracies.

Ten sonic underwater buoys were laid, using radio navigation, to provide accurate reference points for the start of each new main channel; HDMLs were fitted to locate these buoys on the night of the assault, and mark them for the leading sweepers. Then, the dans for the assault swept channels needed to have lights. These were specially developed by '*Vernon*' at Portsmouth, and 1,500 lanterns, plus 200 more with a flashing light, the latter to mark the ends of channels, were produced. On the night, they all worked well and gave a fairyland look to the whole area of sea between the Isle of Wight and the beaches.

Advice to friendly warships

This was in fact a pretty important item since, with a vast number of ships milling about in restricted waters, the sweepers' tails would easily get cut or damaged. Each ship in the invasion fleet, therefore, received tables giving the width and length of sweeps, and time taken by the sweepers to turn at the end of laps. At night all sweeps carried lights on the floats at the end, to warn off friendly warships.

Training

Incredibly, this important item gave a lot of trouble, since it was impossible to assemble the fleet of sweepers away from their normal operational duties far in advance of the night of the assault.

The Canadian fleet sweepers arrived in the UK early in April, but they had previously been employed on anti-submarine operations in the North Atlantic, and only re-embarked their sweeping gear at short notice, before sailing for Europe. They had one month in which to work up, and had a busy time. Winches failed, ropes got twisted round the screws and so on, but they went on to do splendid work in the assault and thereafter.

The American YMSs arrived in May. They had mainly been used before as towboats, so they needed an intensive training period. *Chimo,* the US headquarters ships, only arrived from America on June 5.

The minesweeping plan

The plan fell into five parts, in the initial assault period. The first part was to sweep a new, large circle of water a few miles south of the Isle of Wight. The existing South Coast swept channels would be linked to this new area (which inevitably became known as 'Piccadilly Circus'), and from it ten new swept channels were to be cut south to the landing beaches on the far shore.

The new circle of swept water was cleared by the sweepers with great secrecy some days before D-day, and the swept channels from the assault assembly areas were checked again and again. The Admiralty had been given much food for thought by a recent German air minelaying attack on the Needles Channel, leading out of the west end of the Solent, the biggest assembly anchorage of all.

This raid took place on the night of April 27, but due to the number of sweeping passes required, as the mines had delay mechanisms, the channel could not be reopened until May 1. A wreck in this channel would also be of considerable danger to the British Force 'G', which would leave through there; and the situation was further complicated because fleet sweepers were unable to sweep this channel, due to their draught.

Next, the ten new swept channels leading to the assault area had to be cut on the night before the invasion. These new channels collectively became known as 'the spout' and, for some months, must have been one of the busiest shipping routes in the world.

The Allies knew that the Germans had laid a new mine barrage in mid-Channel, stretching all the way from Cherbourg to the Dover Straits. These mines, which were being laid in the early spring of 1944, were thought to be of the contact variety, and thus a wire sweeping job.

Steps had been taken to impede the German minelaying operation. Eighteen motor torpedo boats from Portland and Newhaven had attacked the minelayers, which were working by night, and during the periods of full Moon. Ten Albacore and 12 Typhoon aircraft made strikes against them as well.

The Germans, naturally, did not want to be restricted themselves by this minefield, and so the mines were set to sink and thus sterilise themselves by June, but inevitably there were some which refused to sink and they did some damage.

In cutting these new channels, the sweepers met a problem all their own; for as they swept southwards during the night, the strong Channel tides would be running first eastward, and then westward, right across their path. So they had to change their sweeps over as the tide turned, in the dark, and in unswept water; they also had to avoid getting too close to the French coast in the very early light of the midsummer morning, to avoid alerting the Germans, and getting shelled themselves.

A special operation was devised to beat this new problem. The fleet sweepers half way over, as the tide turned, would get in their port sweeps in succession, and form a single line ahead. The Senior Officer's ship in front would keep her sweep out, and the little MLs would sweep ahead of her. Then they would turn in succession, and sweep themselves back into the newly cleared channel, waste some time steering northwards, then turn south again, stream their starboard sweeps, and continue cutting the new channel from where they had left off. During this operation the slow landing-craft flotillas following astern of them, and plunging in the short seas, would catch up some of the distance between them so that they would all arrive in the proper formation at dawn off the French coast.

Then as the beaches drew near, they would split off; some would sweep short channels inshore for the bombardment ships, which were following close on their heels. Others would sweep boat lanes right inshore ahead of the assault landing craft, while yet others would sweep lateral channels north of the beaches, for the larger landing ships and supply ships.

Then, again, they would work to widen these isolated areas of new swept water, until the entire assault area off the beaches had been swept. After that they would carry out maintenance sweeps of all these new cleared areas every day.

At night, the sweepers would have a further and different role to play in the defence line which would be established offshore north of the beaches. In this defence line, the fleet sweepers would be anchored nine cables apart, and about six miles from the French coast. They would be in two separate lines and would remain there all night from half an hour after sunset. All ships to seaward of them would be regarded as hostile; they would keep an anti-submarine watch also, by Asdic transmission, and the motor launches would be secured alongside them, ready to move off as required.

The in-fighting that took place on this defence line over a period of months can be appreciated, when one remembers that three ships of the '*Catherine*' fleet sweeper class were sunk on it in a period of one week, by German small battle units.

The coastal sweepers also had night duties. They were to anchor in the assault area in carefully defined positions and keep a look-out for the telltale splashes of mines hitting the water as they were laid from aircraft.

All of these operations were broken down into over 100 different 'serials', and each was to be carried out accurately to a very tight timetable. Basically, the fleet sweepers were to be used for wire sweeping and the coastal sweepers for influence sweeping; but the fleet sweepers were soon engaged in influence sweeping also, and the BYMSs, YMSs, and some of the MMSs too, undertook wire sweeping on a wide scale.

The fleet sweepers get ready

A clear sense of purpose and of destiny comes through in the operations orders of the flotillas for Normandy. Here is a sample from the First Minesweeping Flotilla, *Harrier* Senior Officer: 'The clearance of Channel 9 is to be continued to the lowering position regardless of enemy interference or casualties. In the event of enemy attack, the sweeping formation must be preserved and the attack fought off with the best available means. No ship must be allowed to sink in the swept channel.' The Senior Officer of this flotilla designated who should take over from him, allowing for his own ship and three others to become casualties!

They would be sweeping in 'G' formation to port, with 300 fathoms of sweep wire veered. Two minesweeping MLs would sweep in column, ahead of the Senior Officer's ship, and the dan-buoy wires would be 75 fathoms long.

The orders for the Fifteenth Flotilla read alike. These flotillas were out on the eastern flank, with German naval forces in Le Havre just opposite them; so fleet destroyers were attached to them to protect them, *Scorpion* to the First, *Serapis* and *Scourge* to the Fifteenth. Six motor torpedo boats were with the Fifteenth disposed around the port beam and bow of the formation, while the big destroyers were stationed two cables astern of the danlayers in the rear. One HDML, 1415, was attached to the flotilla to find and point out the sonic beacons already laid. *Fraserburgh* and *Calvay* would be running taut wire measuring checks.

The flavour of the assault comes through in the extra orders for the Fifteenth Flotilla. For example, no gunfire unless ordered, no radar to be operated to the end of nautical twilight, and RCM gear to be switched on seven hours before the landings. Ships were warned that passing Spitfires might jettison their belly fuel tanks, and these should not be mistaken for falling mines when they hit the water!

The orders for the night defence line were also stirring. Steam would be at immediate notice when at anchor, and ships were to guard against possible poison gas when the wind was offshore. The MLs would lie alongside the fleet sweepers and would move off and lay a dan, if any mine were seen to fall.

Orders for fleet sweepers in the American sector followed the same pattern. The Fourth Flotilla, of the old 'Town' Class, was to sweep in 'G' formation, with wire sweeps out to both port and starboard. They were expecting trouble, in the form of E- or W-boat activity, enemy aircraft, glider bombs, mine obstructors of all types, destroyers, U-boats and pilotless aircraft. Only the mines would be routine.

The Fourth Flotilla had four danlayers, but one of these was the Canadian fleet sweeper *Thunder,* and she was to be ready to sweep or lay dans at short notice, depending on casualties. Again, an HDML would mark the buoy positions at the exits and entrances. The starboard terminal dan would carry a red light, occulting every six seconds, and Flag 4 International; dans from there onwards would carry the same flag, but a fixed red light. All dans would be one mile apart.

All ships would keep an accurate plot of the channel swept and the buoys laid. All four ships fitted with the taut wire gear would stream it. Again, 300 fathoms of sweep wire would be veered, with seven fathoms of float wire and 13 fathoms of kite wire; all sweeps would be fully armed with explosive cutters.

In the American area, Y.1 Squadron of YMSs was to follow close astern of the Sixteenth Flotilla, and stream their starboard Oropesa sweeps when close inshore. Astern of them would come the British 132nd Flotilla of MMSs, which had Mark V Oropesa specially fitted for the invasion. They would stream their sweeps in the boat channel, and clear it to inside one mile from the beach, while Y.1 Squadron turned and cleared a fire support channel, two miles offshore from Grandcamp.

The assault sweep

So D-day came, on June 6 1944, and all the complicated minesweeping serials were tested out in practice. For the sweepers, there was action aplenty, but in some ways the assault itself was for them hard, tense work. Only in the American sector were there any serious incidents to start with. But the lack of mines, at least to begin with, in the numbers that had been expected took nothing away from the great achievement of all the minesweepers, who carried out their biggest operation of all time. Not the least of their achievements was in carrying out their sweeping serials with cross tides of great strength, and in rough seas.

British sector

Here, mines were found in mid-Channel, as expected, in the newly-laid German field, but the numbers were not great. The assault report of the Ninth Flotilla, of 'Bangors', is typical. They destroyed about a dozen moored mines, including two swept by the MLs in the van. Two ships, *Tenby* and *Bangor*, found mines in their sweeps when recovering them, and bravely steamed straight out into unswept water, and cut their sweeps. No sweeps were parted, and no obstructors found. The M/S MLs were found to be invaluable, and the standard of danlaying in the dark was first class, with no failures out of the 81 dans laid.

The 150th Flotilla of BYMSs also had a fairly uneventful assault. They sailed from Yarmouth Roads at midnight on June 5–6, and followed astern of the Sixth Fleet M/S Flotilla. At 0230 hours, they streamed their wire sweeps in 'G' formation, to widen the new swept channel, but found no mines. At 0430 hours, they recovered their sweeps and waited for daylight and for the bombardment fleet destroyers they were to sweep in to a firing position just off the coast. As they waited, they could see large fires burning ashore, and anti-aircraft fire on the coast. *Grenville*, one of the fleet destroyers, joined up at 0600 hours, and BYMS 2002 swept her in with double Oropesa, to her firing position 2½ miles offshore. She swept round the destroyer, then closer inshore to within half a mile of the beach, where sweeping was stopped as the water was too shallow.

By this time, shells from the shore batteries were falling all round the sweeper, and several fell between her and her Oropesa floats. At 0955 hours she retired and rejoined the flotilla for close inshore sweeping in company.

The 115th Flotilla of short MMSs also had a fairly uneventful time, though not as peaceful as their report suggests. This flotilla commented that when sweeping off the beaches at night, they flew radar-reflecting balloons from the Oropesa floats of their wire sweeps, and that this ruse was successful in attracting German shellfire!

The American sector

Here, on D-day and the days after, mine warfare activity was more serious than in the British sector, mainly due to a field of magnetic mines laid off the Cotentin peninsula before the landings. This had not been anticipated, hence the emphasis on wire sweeping for contact mines in the opening serials.

For the American sweepers, Normandy began before D-day. The British Fourteenth Flotilla of fleet sweepers, carrying out an Oropesa sweep on June 4 in rough weather south of the Isle of Wight, found a moored field. Sweep Unit 3, composed of the American AMs, was following up astern and, that evening, *Osprey* hit one of these mines. Heavily damaged and on fire, *Osprey* soon went down, her survivors being rescued by her sister ship *Chickadee*.

Then the sweepers for the American sector turned south, to sweep the new channels to Utah and Omaha beaches. The fleet sweepers went first—British, American, and Canadian, using first a double Oropesa wire sweep, later changing to single. But no mines were encountered by this force during the cross-Channel sweep—their main problem was in maintaining station and sweeping an accurate path, in the dark, and with a rough sea churning around them.

Behind them came the two YMS squadrons, Y.1 and Y.2, ready to sweep the inshore lanes on arrival. They, too, had a rough passage, but at 0100 hours on the morning of D-day, about 15 miles north of the coast, they started clearing the assault boat lanes and the firing positions for the bombardment ships. Here, the flotillas had differing experiences. The Canadian Thirty-first Flotilla completed their initial sweep at 0515 hours, and although they had been within 1½ miles of the French shore at 0300 hours, they had received no enemy fire. The British Fourteenth Flotilla, however, came under radar-controlled shellfire from the St Marcouf Islands, so they switched on their radio counter-measure gear and the firing magically stopped.

The coastal sweepers, too, had mixed times. Those for Omaha beach had an easy time; the water was deeper and so the sweeping was easier. Although the British 104th Flotilla of MMSs went within one mile of the beach, while sweeping with LL and SA for the boat channel, and leaving lighted dans behind them, they were not molested. The YMS squadrons, too, passed within a mile of the St Marcouf Islands, but met with no resistance.

However, those approaching Utah beach had a different story to tell. As the squadron was changing its sweeps, the ships came under heavy fire from shore batteries at St Vaast. Two British cruisers, *Black Prince* and *Hawkins*, came tearing up at high speed and carried out counter-battery fire to protect the sweepers. But for a while, the situation became quite hectic; YMS 380 had her Oropesa sweep cut by a shell, and a friendly plane which flew down the shoreline to lay protective smoke was hit and blew up in front of the sweepers.

Yet the British 132nd Flotilla of MMSs had the opposite experience. They streamed their newly-fitted Oropesa sweeps astern of Y.1 Squadron, and swept in to within one mile of the beach, within small arms range, laying lighted dans for the assault boats. They even turned 180° right under the noses of the enemy, and came back over Cardonnet Bank, a shoal that was to prove troublesome later that day, but still no enemy fire engulfed them.

By this time, the assault was under way on the beaches, and the minesweepers turned to their next task of sweeping lateral channels off the beaches, and widening the swept area to include most of the unloading area.

Soon, mines began to claim victims, mostly from the magnetic field laid just before D-day on Cardonnet Bank, off the St Marcouf Islands. This field was to cause many casualties, and some disruption of the assault, before it was finally cleared. The US destroyer *Corry* was the first victim, only three minutes before the assault hit the beaches; she was hit by a mine some two miles off the beach and sank quickly. Next, a US Navy PC, 1261, was sunk, about 2½ miles off the beach and, in the course of the day, 16 landing craft were sunk by mines in this sector.

The sweepers were very active all day off Omaha beach, sweeping and picking up survivors. The US 'A' Squadron, of the 'Raven' Class AMs, had a roving commission in the assault area, using all three types of sweep, wire, magnetic, and acoustic, and the YMSs were equally busy.

The British 132nd Flotilla of MMSs, also operating in this area, but closer inshore to the beaches, had a different day—plenty of action, but no mines. They carried out an LL/SA sweep inshore of the five-fathom depth line, and were almost continuously under shellfire for their trouble. Some 60 rounds fell within 500 yards of MMS 261, the Senior Officer's ship, but though shrapnel pitted their sides, and rattled across their decks, none of the ships were hit. The Senior Officer's ship, however, leading the flotilla in shallow water, went aground in the middle of all this, and though she got herself off again in short order, her towed SA sweep box was damaged. It is an interesting comment on the shortage of equipment that ships in this flotilla were not fitted with depth recorders.

The Canadian Thirty-first Flotilla of 'Bangors', also in this Omaha area, found no mines that day after sweeping with Oropesa for some 18 hours. This flotilla, like the others, averaged four hours' rest each night, but got 78 mines in their first seven days. The 167th Flotilla, of BYMSs, and the 104th Flotilla, of MMSs, sweeping with LL and SA, equally lifted no mines during the day. After sweeping all the previous night, down from England, they anchored only at midnight, and then kept a watch for mines splashing down from aircraft.

Next day, June 7, the trouble on Cardonnet Bank went on. Most serious was the loss of the American fleet sweeper *Tide*, when she, with *Threat*, *Swift*, and *Pheasant*, was searching the area for mines believed laid from German E-boats during the night. *Tide* was struck in the morning, four miles from the St Marcouf Islands, and suffered 115 casualties. *Pheasant*, *Threat*, and PT 509 moved in to rescue the survivors, and the resulting photograph is one of the most moving minesweeping scenes of the war. It was thought that *Tide* was inside the minimum safe depth against magnetic mines for her class when she was struck.

The fleet destroyer *Swift*, of the war-built British 'S' Class, was also sunk by a mine in the area that day, together with three more landing craft, while YMS 406 was damaged by a near miss. But during the day, 30 mines were detonated near the boat lanes to Utah beach, so the clearance of this important field was started, even though the Germans replenished it by aircraft at night.

June 8 saw a continuation of this mine warfare battle. The US destroyer *Glennon*, on bombardment duties, was sunk at eight in the morning, and while *Staff*, one of the AMs, was assisting her, the destroyer escort *Rich*, trying to help as well, ran over three of these mines in just a few minutes and sank quickly with 152 casualties.

Then another US destroyer, *Meredith*, was bagged by a magnetic mine in the same area, and the British netlayer *Minster* was also sunk with 70 casualties, as her crew were below decks in the lunch hour. Another US destroyer, *Harding*, was damaged, and LST 499 and some more smaller landing craft were sunk. That field was proving to be very effective.

The following day, June 9, the sweeping continued, with the American AMs actively involved. Seven ground mines were soon lifted off Quineville, but the sweepers were under fire from the shore batteries at St Vaast. A total of 17 ships was damaged that day, including six of the valuable LSTs and a big supply ship. When fired on from the shore, the AMs replied with their 3-inch 50-calibre guns, and *Threat* and *Swift* claimed several hits.

The battle between the AMs and the shore batteries continued over the next few days. The sweepers spotted inshore, while still sweeping, for the larger bombardment ships operating further out. *Staff* lost her magnetic tail to a shell, and *Chickadee* received heavy shrapnel damage. Sometimes the sweepers had to retire at speed when the shells got too close, but clearance work continued, though assault traffic had to be diverted away from Cardonnet Bank. Ships continued to be sunk or damaged there for quite a period.

The build-up—mid-June to September 1944

For the minesweepers, the battle did not end with the success of the original assault; quite the reverse. Even allowing for the difficulties with the Cardonnet Bank field in the American sector in the first week, the Germans mounted an intensive mining campaign against the beach-head over the next few months, until other French ports, and Antwerp, were liberated and cleared and the Normandy beach-head was closed down.

The 'Oysters' appear

The German campaign used all three of the mine types already in use—moored contact, magnetic, and acoustic—but they also introduced for the first time their pressure mine, which the Allies code-named the 'Oyster'. This mine was ready for use by the Germans in 1943, but in view of the Allies' success in recovering specimens of the magnetic and acoustic mines soon after they were introduced, and devising an antidote, the Germans held the pressure mines in reserve for the impending Allied invasion, to cause the maximum disruption at that time.

By May, the Germans had a good stock of these mines ready—pressure, acoustic-pressure, and magnetic-pressure variations—including 2,000 stored in underground hangars at the airfield at Le Mans, with two squadrons of minelaying aircraft standing by to lay them in the assault area, whenever the invasion started.

In the first six weeks after D-day, over 400 of these 'oyster' mines were laid, 216 of them between June 11 and 14 alone, in the Seine Bay area. They disrupted the assault traffic quite seriously. Indeed, it is surprising that the Germans did not use more of these mines against the invasion, for they turned out to be virtually unsweepable, and drastic measures had to be taken to minimise their effect in the assault area.

Experiments had been carried out for some years by the Royal Navy, and now all the early hard work on ships' signatures and swell movements paid off. From the records kept in the United Kingdom since 1941, a table of safe speeds was sent to the Normandy beach-head within 48 hours of the first pressure mine being identified. All ships in the shallow assault area of the Seine Bay were instructed not to exceed these speeds, and the worrying losses of ships were almost immediately reduced in proportion.

The first speed restrictions looked like this, and the effect on crowded invasion traffic can be imagined:

Depth of water in fathoms	**5**	**10**	**15**	**20**
Safe speed, in knots				
Battleships	0	2½	4	5–6
Destroyers	3½	7	11	14
'Liberty' ships	0	4	6	9

Then pressure mines began to be recovered and, as with the magnetic and acoustic mines earlier in the war, dissection of recovered mines was a turning-point in finding effective sweeps against them.

The first two mines, laid by aircraft at night and landing ashore, were recovered at Luc sur Mer, in the British sector, on June 22, swiftly followed by a third the following night at La Croix, and a fourth was found by the US Navy on June 30. Four more were recovered ashore in the next ten days, at Courseulles, Lion sur Mer, and Langrune; and on July 25, one was found floating in the British assault area, partially exploded, but with its tail and pressure unit compartments intact. This was quickly flown to England for examination, as those found on land usually had their relays damaged by the explosions of bombs and shells landing near by, and so were difficult to analyse.

The Royal Navy's scientists found that in the pressure unit a rubber bag acted as an air reservoir and served to communicate the water pressure to a thin aluminium diaphragm which separated it from the lower air chamber. When the water suction pulled the bag away from the diaphragm, a switch was closed, actuating the mine. A small leak was arranged, so that the water pressure could slowly equalise inside the unit, and it stabilised itself in any depth of water, so that only sudden reductions in pressure over a period of about ten seconds would cause the switch to be closed.

The pressure mines were usually combined with a magnetic or acoustic unit and, in the first mines recovered, it was found that a reduction of 1½ inches on the pressure side persisting for eight seconds would actuate the pressure unit's switch, and thus make the acoustic side live, and the mine would fire if an acoustic signature of sufficient intensity was received within 30 seconds of this happening. If no signature was received, then the mine returned to normal. These first mines were designated by the Germans Type AD.104, and by the British Type AP Mark I.

Living with unsweepable mines

The speed restrictions so quickly introduced into the Seine Bay area had an immediate effect on the casualty rate; but the Allied had something else going for them—the weather. 1944 turned out to be a stormy summer off Normandy; the great storm wrecked the Americans' Mulberry harbour, and severely damaged the British one, with great losses of valuable landing craft. But the same storm whipped up high seas, and after them a heavy swell. This massive movement of water actuated the pressure side of the 'oysters', so that the sweepers were able to clear them as plain acoustic mines. In the Seine Bay especially, the sweepers used this knowledge to good effect, even closing the assault area to all movements at favourable times of swell and tide, so that they could carry out an intensive acoustic clearance operation.

It was lucky that the British had done so much work on swell recording in the earlier war years; now, a swell-recording unit which they had invented was laid on the bottom in the Seine Bay to help the sweepers in their work. Soon similar units were laid at strategic points off the South and East Coasts of England, and later off the continental coast, from Cherbourg to Flushing. They were able to forecast when acoustic sweeping was likely to be effective against the 'oysters', and great care was necessary, since a small swell only made the pressure units of the mines more, and not less, sensitive. But with a heavy swell of two inches or more of suction, the pressure unit's switch was permanently closed so that the mine just became a plain acoustic.

The wave-making sweeps return

On June 23, the First Sea Lord said in London that safe speeds were the only known answer to the 'oysters', and ordered further urgent wave-making trials with coastal forces units. *Cyrus* was repaired and towed to the Thames, for use in the Estuary, if needed.

The seriousness of the threat posed to Allied shipping by the 'oysters' is shown by the use of the Royal Navy's latest and best coastal forces units in these experiments. Not only were some of the latest 'D' Type MTBs used, but also the one entire flotilla of six steam gunboats (SGBs); the latter were particularly useful as they were much bigger and could carry the heavy minesweeping gear at greater speeds.

The first new experimental sweep used three 'D' type MTBs, towing a sound source as an SA sweep. This source was, in fact, the pipe noisemaker developed for towing by frigates and corvettes against German acoustic torpedoes in the North Atlantic; designated Type E, it was towed 400 yards astern by the centre boat in a formation of three. This gear was then known as SA Type D Mark I. In a parallel trial at Weymouth, three MGBs (46, 13, and 72) carried out the same trial but with a station-keeping distance of only 15 yards, in arrowhead formation.

Sometimes an SGB was used as the centre boat and, in calm water and in depths up to 12 fathoms, they swept the 'oyster' neatly over a path two cables wide. It was hoped the mines would explode 300 yards astern of them. However, if a swell was present then the sweeping MTBs were themselves endangered, as the mine, by that time turned into a plain acoustic by the swell, would be fired by the combination of the first waves of the boats' wave train, at it was called, and the high-pitched noise of their fast-running propellers.

No SA gear had been developed for ships operating at that high speed, so that they were not often used off Normandy, since the weather produced almost continuous swells there that summer. But if the Germans had dropped large numbers of acoustic 'oysters' in the shallow Thames Estuary that year, these wave-making MTB sweepers might have proved their worth.

By mid-July 1944, the Allies suspected that the Germans were introducing a magnetic-pressure mine into the Normandy beach-head, and took energetic measures to combat it. A combination of MTBs as wave-makers, and the well-tried LL magnetic sweeps seemed to be the answer; but the gear was bulky and heavy, and even the large MTBs found difficulty in varying it. So the First SGB Flotilla was converted to LL sweepers in double-quick time. They were fitted with special tails—for the centre ship of three, an LL tail, with the long leg 500 yards long, and the short leg 100 yards, operating at 750 amps. This gave a towing strain of one ton, which was acceptable, and a 75 kW petrol engine was installed to energise it. For the two wing ships, a single L tail was provided, energised by a 20 kW generator in each ship. They swept at 20 knots or more, and swept a path 200 yards wide. An HDML was attached for swell-recording duties, and the gunboats operated 50 yards apart, with only the centre ship operating her acoustic sweep as well. Dan-buoys were laid by the outer boats from their sterns.

Then, on July 24, the first magnetic-pressure mine was recovered ashore. Designated the DM.1, it was found to operate quite differently from its acoustic predecessor. The magnetic side was a standard five milligauss unit, while the pressure side was actuated by a reduction in pressure of 1½ inches, present continuously for seven to nine seconds. It was not actuated by swell, nor by the SGB wavemakers, so the SGB LL sweep was found to be ineffective.

Experiences in the two main sectors were remarkably similar—in the month from the assault to July 3, 291 mines were swept in the British assault area, and 261 in the American sector. Laying and sweeping went on for some months after that.

The eastern (British) sector

After the initial assault, the British minesweeping forces settled down, not to routine sweeping, but to a hard slog carried out under difficult conditions and with a fast increasing number of mines being swept.

Figures are often conflicting, but while records show that 291 mines of all types had been swept in the British sector between June 6 and July 4, that total swelled in the following 2½ weeks to 569; and that figure was in turn made up of 97 moored contact mines, 124 magnetic, 216 acoustic, and 132 unknown. Note that pressure mines were not being identified separately at that time.

Further, it was estimated that by July 4, about 600 new mines of all types had been laid in the British assault area since the initial landings. It was thought that about half of that new total had been swept, or had detonated themselves, for the loss of 14 ships sunk or damaged.

The sweepers worked hard all day, with routine sweeping of the ten approach channels, and of the lateral assault area. These sweeps had to allow for delayed arming of the mines, and for ship-count mechanisms; so that in no case was a single pass by the sweepers considered safe. A further hazard was the mass of shipping of all types, each ship intent on making its contribution to the build-up of forces and supplies put ashore in this critical period. So landing craft were heading to or from the beaches, across the sweepers' path. Supply ships and big bombardment units were moving slowly into their anchorage positions; escorts, coastal craft and tugs were moving in all directions; and the sweepers, with their precious tails out, had to avoid them all. But there were plenty of risky moments, and many a wire sweep or LL tail was cut. The vast majority of the crews of the other ships had no experience of working closely with sweepers and did not appreciate the length of their tails, nor their inability to manoeuvre while sweeping.

There were other hazards, too. Sweeping in these shallow waters, with an increasing number of wrecks strewn around, made for parted sweeps. The amount of flotsam in the area was a real menace, in fouling sweeps and propellers—so much so, that a special garbage hunt was organised daily among the escorts, with candy for prizes! The Germans were also aware of this situation and, on a number of occasions, floated mines down to the anchorage on the tide.

Another problem was that the anchored shipping, massed but not in formation, prevented the sweepers maintaining their usual 'Q' or 'P' formations and, on many occasions, single-ship sweeping was necessary to cover the area.

The BYMS and MMS flotillas had other problems to cope with; the 156th Flotilla, of BYMSs, carried out three periods of sweeping in June, bagging 60 mines. Only a week after D-day, they began sweeping an area towards the mouth of the River Seine, which was still occupied by the Germans, so the fleet destroyers *Jervis* and *Sioux* (the latter Canadian) went along as escorts. Five ships were sweeping with double Oropesa and two were laying dan-buoys. Shortly after starting, three acoustic mines exploded close astern of BYMS 2003, and she was badly shaken up; but their minds were soon taken off that by accurate shellfire from the shore, at a range of 14,000 yards. They were straddled by 5.9-inch shells in the first three salvoes, and retired without damage, under cover of smoke. Next day, they swept the area again for magnetic and acoustic mines, and lifted seven magnetics. This time they were shelled again, from Ouistreham, but completed their sweep.

Their next sweeping period coincided with the very bad weather, and the work was found to be difficult. Then, when they anchored they dragged their anchors, and four were lost. But in spite of the weather, on June 28 they lifted six magnetic and 16 acoustic mines; the latter all exploded in a period of four minutes, detonating in series of three at a time. Several mines went off very close to the sweepers, but none were damaged.

There was little rest for the smaller sweepers at night; anchored in the assault area, they had to keep watch for falling mines, and catch what rest they could, as well as sharing out their precious stores. They commented, perhaps reasonably, that as they had one small boat each, with no engine to drive it, the ships had to come alongside each other for orders and were sometimes damaged in the process.

Impromptu repairs were the order of the day. MMS 27, of the 101st Flotilla, got a wire rope round her propeller, and promptly beached herself in order to clear it. Not only was this successful, she scraped and cleaned her bottom at low water into the bargain, and her example was thereafter copied by her sisters!

July

Enemy minelaying by aircraft at night continued on a number of occasions; about 100 mines were laid in this way in three weeks, rather less than the average since D-day. In addition, 15 mines were seen to explode as they hit the water. The German aircraft were observed to be approaching the anchorage at low altitude from landward, so barrage balloons were flown in the Juno beach area at operational height at night and not brought down, as had been the practice.

Great care was taken to sweep daily the area occupied by the outer night defence line of fleet sweepers, since this early warning and defence line was essential for the security of the fleet. Each day this area was swept 20 times, to cover actuations on any mines present, and this gives an idea of the stress under which the sweepers were working.

A plot was being kept, showing accurately the positions of all mines seen to fall, together with the age of each mine. An intensive sweep was carried out seven days after each mine was laid, to catch it as it became ripe; and the explosive sweep, Mark I, with single ship LL/SA, was used successfully; 11 acoustic mines were swept by this method in one week, and BYMS 2074 swept a further ten in one day.

On one occasion, a mine was laid very close to the bombardment cruiser *Hawkins*. This mine received very special attention, but eventually yielded to treatment 11 days after being laid. It detonated on the 52nd sweep over it, much to the relief of *Hawkins*; a relief not reduced by the sweepers' comment that it had almost certainly been an 'oyster', and a slight swell rising had been the only reason for their success.

Daily searching of the swept channels was being carried out continuously with the sweepers available, but their total, large though it was, was only just sufficient to achieve this. Maintenance was becoming a real problem, and strenuous efforts were made to keep the sweepers going. *Ambitious,* the headquarters ship, moved inside the Arromanches Mulberry harbour, which gave greater calm than her old anchorage off Juno beach, and there two BYMSs or MMSs could berth on either side of her for repairs. The BYMSs were proving their worth in the British assault area, and were standing up to the strain better than the MMSs; the average daily sweeping strength in the 102nd, 115th, and 143rd Flotillas, all of MMSs, was down to five ships each, and the ships were no longer standing the pace. The fleet minesweeping flotillas were averaging no more than four ships per flotilla. But morale and determination stayed high, and when the sweepers got a special performance from the first concert party to visit the beaches, they had an enthusiastic audience.

It was the proud boast of the sweepers that no warship or merchant ship was a mine casualty in the British assault area in this period; but they themselves were not unscathed. For example, BYMS 2252 was damaged by fouling the anchor cable of another ship; BYMS 2079 was towed home after hitting a rock, and only two days after being badly shaken up by the near miss of a mine; BYMS 2016 also suffered damage in a near miss; MMS 205 developed a serious engine defect, and had to be towed home, while MMS 183 produced bad defects in her LL batteries. The fleet sweeper *Chamois* caught a near miss, and had to return to Plymouth, as did MMS 102. And so it went on—but so did the build-up over the beaches, and that was all that really mattered. Even the high number of mines now being laid was not holding up the flow, through the devoted toil of the sweepers. But the hazards were great; MMS 55 was blown up while carrying out an acoustic sweep, and it was decided on several occasions to suspend sweeping until a swell had moderated.

In this period, they bagged 64 magnetic, 40 acoustic, and 42 unknown, for a total of 146; note that no moored contact mines were now included in the number.

August

This month produced a lull in the rate of lay of new mines, but many still remained in the anchorages and the speed of all ships was still severely restricted as a defence against the pressure mines. To show the care being taken, just ten mines were dropped by aircraft on two nights, in two channels; some of the mines dropping were even plotted by radar. These two channels were swept 49 and 53 times respectively, in a period of 12 days, working off it was hoped, all the pulse delay mechanisms; though it was feared that the magnetic-pressure mines had still not been neutralised and were lying there in wait.

On mornings when there was a swell, which helped to set the 'oysters' off, traffic in the assault area was stopped from 0630 hours to 1000 hours, while all available sweepers made a concerted attack on the mines, using extended formation to cover a wide area. This was found to be successful. Then a BYMS flotilla spent two days extending one of the channels to a mile wide, by an intensive sweep carried out 22 times, and lifting ten magnetic mines.

Other special activities included the sweeping of an approach channel to the little port of Ouistreham, now liberated. BYMSs first carried out an Oropesa sweep, then MMSs followed with LL/SA. Then portable pulsers were used to sweep the inner area of the harbour. From there, the sweeping of the canal inland to Caen commenced. For this, an 'A' sweep was used (a rare occurrence in this war), with three-ton trucks running along the towpath (flying White Ensigns, of course), and preceded by two MFVs fitted with an LAA sweep. Much of this sweep was carried out under shellfire.

During the month, about 60 mines were laid from German aircraft and 130 were swept, so at last the balance was going in the right way. Of this last total, only three were moored contact mines, 71 were magnetic, 16 were acoustic, and 40 were unknown. The last week of the month alone produced a mine bag of 75, the highest for many weeks.

The fleet sweepers were being used now for Oropesa clearance of special areas, but they too were following up with LL/SA sweeps, as the smaller sweepers were fully stretched inshore.

There were some casualties during the month among the assembled shipping. If they were hit while in the swept channels, it was thought that they were the victims of pressure mines, and had been going faster than their safe speeds.

Surprisingly, shellfire from German batteries ashore was still persisting. The 150th Flotilla, of BYMSs, was shelled very accurately on August 20, near Trouville. Conversely, the German small battle units were now operating around the assault anchorage from time to time and explosive motor boats, midget submarines, and human torpedoes were warded off by the sweepers both by day and by night. BYMS 2035, for example, came across a human torpedo, took 'vigorous offensive action', and reported that the crew was only too willing to surrender to the sweeper!

However, casualties among the sweepers inevitably continued to mount. MMS 165 was damaged by two near misses, MMS 29 grounded while sweeping, and MMS 49 produced a serious engine defect. The fleet sweeper *Vestal* (an 'Algerine') was damaged by a near miss from an acoustic mine on August 10, near the light vessel now established in the Juno area, and another fleet sweeper, *Gleaner* (a 'Halcyon') was similarly damaged.

A weird and motley crew arrived in the British assault area during the month. The Mine Recovery Flotilla, from *Vernon*, the mining school at Portsmouth, came to recover any interesting mines they could find, for dissection. They went to the British Mulberry Harbour on June 27 to recover 'oysters', and on to Cherbourg on July 14 to recover a German tripod mine, of a type causing much difficulty. But no brand new warships here—the flotilla consisted of two old motor yachts, *Esmeralda* and *Brinnaric*, fitted with Asdic and echo-sounder (thus were the minehunters of today spawned), and two equally ancient drifters, rejoicing in the names of *Fisher Boy* and *Young Cliff.*

September

Another big bag of mines was produced, though no new ones were seen to be laid during the month. The Germans' attention was already turning to the shipping lane between England and Antwerp, and the Normandy beaches no longer held the attention of the world. But the month's bag totalled 166 just the same, made up of 133 magnetic, 20 acoustic, just three moored contact, and 14 unknown.

Near misses on sweepers continued (MMS 49 caught one only ten yards ahead of her from an acoustic on September 5) but no sweepers were lost tc mines during the month. It is good to note the excellent record of the British assault area sweepers, in not having ships sunk on their patch, in spite of the great mine warfare campaign which had been conducted there in the past four months.

Following a radio broadcast in England, there was some competition among the sweeping flotillas at this time as to who had swept the most mines in a short period. The 150th Flotilla, of BYMSs, claimed a special feat on July 16, lifting 22 ground mines in four minutes, including six magnetic in 16 seconds. The 165th Flotilla countered by claiming a mixed bag of 23 mines on the same day, while the 163rd Flotilla claimed 24, and all magnetic, on August 31.

At the end of the year, with the other northern European ports liberated and in use by the Allies, other than Dunkirk, the need to use the long truck route from the Normandy beaches fell away, and the British assault area was closed after a short but brilliant history.

The western (American) sector

The mine warfare story in the American sector, after the first few hectic days around the Cardonnet Bank, predictably followed the pattern we have seen elsewhere. The fleet sweepers (especially the British and Canadian flotillas attached) concentrated on daily clearance sweeps of the swept channels leading south from the Isle of Wight, and the deeper water assault anchorages, while the YMSs, BYMSs, and MMSs worked ceaselessly in the shallower inshore areas.

The fleet sweepers did not find too much, except when the Canadian Thirty-first Flotilla located a field of moored contact mines on June 9, and cleared 53 mines, most of which detonated in the sweeps. The following day, *Tadoussac* received a direct hit from a shore battery, and had to return home.

The smaller sweepers also found most of the difficulties encountered by their colleagues in the British area, with the exception that they did not suffer from the small German battle units, mines drifting on the tide, or night attacks, since the Germans launched these mainly from the eastern side of the bay.

They swept all day and anchored at night, in the vicinity of *Chimo*, their headquarters ship, to watch for falling mines. One difficulty they found was in maintaining the buoyed channel marks; on one occasion, a big 'Rhino' ferry was seen with three lighted dans trailing alongside, after it had fouled them, and the LSTs were also said to be offenders in this respect.

After the big field of magnetic mines was found, all sweepers so fitted used their LL tails; frequent night raids were made by enemy aircraft, but LL/SA sweeping of all swept channels was always resumed at daylight. *Chimo* was by now keeping a mine plot, and the danger of diverting sweepers from the swept channels to investigate suspected mine drops was outlined to all.

The 132nd Flotilla of MMSs came under quite a lot of accurate shellfire, and did some invaluable spotting of shore batteries, for the larger counter-battery support ships lying further offshore. Soon, it was possible for two of the YMSs to start laying semi-permanent buoys in the boat lanes, which made life a lot simpler for the landing craft.

In the first week after the assault, 178 mines were swept in the American sector; 71 moored contact, 90 ground, and 17 by ships running over them. Pressure continued after that from the Germans so that, by the end of July, no less than 454 had been cleared, of which 272 were in the beaches area, the rest in or near Cherbourg. The mines near the beaches were made up of 194 ground mines, and 78 moored mines.

Casualties to the sweepers continued to mount. MMS 29 was sunk by an acoustic on Cardonnet Bank on June 13; she was the centre ship of three, sweeping in formation. BYMS 2069 was damaged by an acoustic on June 29, and on July 6 MMS 219 was badly holed when she struck an underwater wreck. The YMSs also took casualties. YMS 279 was damaged by a near miss on June 16, YMS 377 on June 18, and YMS 347 on July 2. Two YMSs, 305 and 377, also had further near misses, as did MMS 297, before the end of the month.

On July 30, a YMS squadron ran into more trouble with ground mines off St Vaast. YMS 304 lifted an acoustic almost directly beneath her, and sank in two minutes; YMS 378 lifted three at once, and suffered a near miss; and the British trawler *Sir Geraint*, acting as danlayer to the squadron, was also damaged by an acoustic mine but safely reached harbour. Mining casualties in the area continued, though at a reduced rate; 13 in the rest of June, and four more in July. But by then, the mining campaign in the American sector had finished.

During the great storm at the end of June, in which the American Mulberry harbour was destroyed, the sweepers in this sector also had trouble in remaining at anchor. Both the AMs and the YMSs parted their cables, and then had to remain continuously under way. The British MMS flotillas attached rode it out at anchor, and the US Navy commented that the ground gear in the YMSs had proved to be too light.

On July 2, Y.2 Squadron of YMSs left for the Cherbourg clearance; they had been preceded by the AM Squadron, which had returned to England a few days earlier, to prepare and rest. But the other YMS Squadron still ran into trouble now and again. Sweeping off St Vaast on July 30, YMS 304 lifted an acoustic mine right underneath her, at nine in the morning; she sank within a minute. Then YMS 378 took a near miss only ten yards ahead and, although towed back to England by YMS 382, she was written off as a constructive total loss.

The rest of the YMSs in the American sector left for Cherbourg during August, but before closing this chapter of great achievement, we should record that on Bastille Day, July 14, a highly successful minesweeping conference was held in the little mayor's parlour at the harbour of Port en Bessin. There were many explosions, but it was reliably reported that these were from champagne corks, and not from mines!

Coastal minesweepers clearing magnetic mines from the Cardonnet Bank, near the American landing beaches at Normandy. This field caused much trouble in the early days of the invasion.

Minesweepers for Normandy—a summary

		British sector	American sector
		(ships and flotillas)	
Fleet minesweepers (by class)			
'Algerine'	RN	25—3	
'Bangor'	RN	16—2	13—1½
'Bangor'	RCN		15—1½
'Catherine'	RN	12—1	
'Raven'	USN		11—1
'Halcyon'	RN	9—1	
'Town'	RN		9—1
	Totals	62—7	48—5
Coastal minesweepers			
BYMSs	RN	30—3	10—1
YMSs	USN		21—2
MMSs	RN	30—3	31—3
	Totals	60—6	62—6
Danlayers (plus 8 Fleets in numbers above)			
War-built trawlers	RN	6	7
Prewar trawlers	RN	19	6
	Totals	25	13
Motor launches (fitted Oropesa)			
'B' Class MLs	RN	20	16
Force sweeper totals		167	139
Total number of sweepers in assault		306	

The American AM Tide *sinking off the Normandy beaches, after striking a magnetic mine in the opening stage of the Normandy invasion. The censor has erased the radar aerials, but a sister ship and PT boat are standing by to rescue survivors.*

Typical fleet minesweeping flotilla at the assault
8 'Algerine' Class fleet minesweepers
1 'Isles' Class fleet danlaying trawler
3 Prewar danlaying trawlers
4 MLs, skim-sweeping ahead

Clearing the French ports

As the Allied armies spread out from the beach-head, so the sweepers also spread out, clearing the French Channel and Biscay ports of mines, as they were liberated by the army. This great operation went on from the beginning of July 1944, right through into 1945. But the very first clearance was the greatest of them all.

Cherbourg

This great port, situated at the northern end of the Cotentin Peninsula, lay just west of the beach-head, and naturally became a great objective for early capture by the Allies, so that its spacious port, with deep water and sheltered anchorages, could handle the bulk of the supplies needed by the invading armies.

Naturally, the Germans defended the port with great energy, and they had also constructed batteries of big guns right along the north shore of the peninsula. But the port was captured by the American Army on June 27 and, although many of the great gun emplacements had not yet been over-run, the sweepers at once started moving in.

They were preceded by a bombardment force of battleships and cruisers, in a great effort to demolish these shore batteries, and they were swept in by two minesweeping units. The first, Sweep Unit 1, comprised the Royal Navy's Ninth Flotilla of eight 'Bangors' (*Sidmouth*, SO), together with four danlayers, four motor launches, and a flotilla of BYMSs. The second consisted of the American AMs, eight of them, with the Canadian 'Bangor' *Thunder* as danlayer, and four motor launches.

An approach channel 28 miles long was cleared from the open sea for the bombardment ships, to bring them in about eight miles off Cap Barfleur, at the eastern tip of the peninsula. Then a fire support channel was cleared laterally for

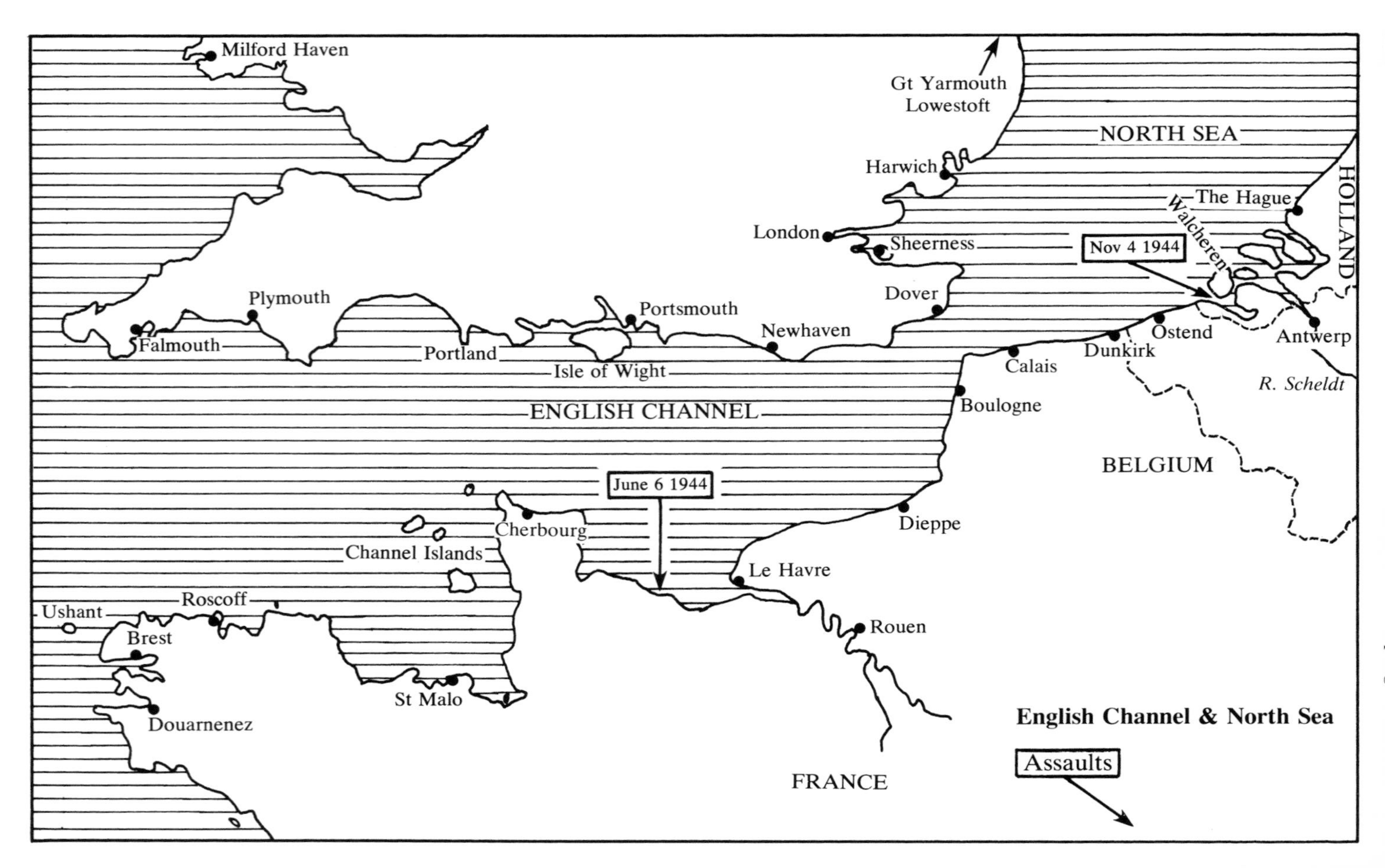
Milford Haven
Gt Yarmouth
Lowestoft
NORTH SEA
Harwich
HOLLAND
The Hague
Walcheren
London
Sheerness
Nov 4 1944
Dover
Plymouth
Portsmouth
Newhaven
Falmouth
Portland
Isle of Wight
Calais
Dunkirk
Ostend
Antwerp
R. Scheldt
Boulogne
ENGLISH CHANNEL
BELGIUM
June 6 1944
Cherbourg
Dieppe
Channel Islands
Le Havre
Ushant
Roscoff
Brest
Rouen
St Malo
Douarnenez
English Channel & North Sea
Assaults
FRANCE

Canadian fleet sweepers clearing the first channel towards Cherbourg in July 1944. Blairmore *is the next ship in line.*

eight miles along the shore. A fierce battle was fought between the bombarding ships and the shore batteries, with honours about even.

The sweepers came under fire, especially when the big ships were hidden by smoke and, after their escorting destroyer took a direct hit, the sweepers were forced to retire northwards at high speed. No sweepers were hit but they were all straddled by shells, while four MLs made smoke around them. One danlaying trawler, built many years before, easily exceeded her designed speed on this retreat! But two days later, the batteries were silenced, and the sweepers began in earnest.

Allied intelligence indicated that the whole of Cherbourg, the outer and inner harbours, the quays and basins, were a mass of mines of every known type, of sunken wrecks, obstacles and booby traps. The sweepers and wreck disposal experts approached the area gingerly, and found the information completely accurate. So, although there was the greatest urgency to clear the port for the invasion shipping, every caution was observed.

The first priority was to clear areas for landing points and anchorages in the vast outer harbour with its great breakwaters. The great sweep started on June 30, with the Ninth (RN) and Thirty-first (RCN) Flotillas sweeping an approach channel through the two main entrances. The first sweep, by the fleet sweepers, was with Oropesa, and this was followed by LL/SA sweeping, 16 passes being made by the 159th and 206th MMS flotillas. In the meantime, the American Y.2 Squadron of YMSs started an Oropesa clearance sweep off the eastern entrance.

Many mines were found in this outer area, and although the wire sweepers operated for only 2½ hours each side of high water, to minimise the risk, several casualties occurred. The British MMS 1019, carrying out an influence sweep on July 2, was sunk by a contact mine with an extended snag line near the surface, just outside the breakwater. The same day, the YMSs ran into a large number of contact mines off the eastern entrance. YMS 350 put up three of these mines in one go, but one of them exploded right underneath her and she sank. YMS 347 caught a very near miss, but was able to continue, and this squadron swept 22 mines in a week.

Clearing the inner harbour

Then the sweepers moved into the Grande Rade, the outer anchorage, and careful clearance commenced inside. The fleet sweepers remained outside and all the inner clearance fell on the YMS, BYMS, and MMS flotillas, with the little LCVPs and divers working in the basins and on the quays. The whole harbour was strewn with wrecks, of every size and shape, and even before all the mines had been lifted the salvage vessels started work. This was primarily an American-directed operation, whereas the sweeping plan was under a Royal Navy officer.

Every type of mine was there and many of them had delayed-action mechanisms, or ship-count devices, so that to be sure an area was clear of these mines, it had to be swept eight times magnetically, and eight times acoustically, every day for 85 days! This was really an intensive mine-clearance operation, although confined to just one large harbour.

In clearing the Grande Rade, MLs took the initial lap, ahead of the YMSs and BYMSs, using Oropesa, and the 159th Flotilla came behind them, though only the MLs could get right into the corners. It was found that the YMSs/BYMSs were the most suitable type of sweeper for this clearance; they were superior to the MMSs in speed, manoeuvreability, draught, and the types of gear that they carried. The ML sweeps were found to be too light, they parted easily, and the MLs were hard to handle when sweeping in a confined space.

In the first two weeks, about 133 mines were cleared, of which 92 were moored contact, 35 magnetic, and two acoustic.

As they moved into the Petite Rade, or inner harbour, the snag line sweepers, the LCVPs, led the way, followed by MLs using wire sweeps, and finally by the 159th Flotilla of MMS, using LL/SA. The MLs cleared away many GZ type mines in this area, but one LCVP was lost in the process. The depth to which they could sweep, and the hours of sweeping possible, were regulated by the tides which have a very high rise and fall here, and by the obstructing wrecks.

The Raz de Bannes area, inside the western entrance, became a notorious place for magnetic mines, and BYMSs, YMSs, and MMSs were all constantly active in this area. No less than 62 'red' mines were lifted here, and 30 on July 23 alone, with 19 going up in six minutes, in series of three. The ships were sweeping in 'Q' formation, and were making 9 knots over the ground at the time.

Meanwhile the MLs continued to find moored mines in the Petite Rade, probably 'Katies', as some had snag lines; but each day, between three and 15 magnetics were put up in the Raz de Bannes area. On July 30 they got their first 'blue' mines here, and some acoustics as well.

Clearance in the inner harbour was spreading to the big dock, the Darse Transatlantique, and the Bassin Charles X near by. An ML was the first to enter the Darse, with a wire sweep out, using 80 fathoms of wire and five fathoms of float wire. Her main hazard was from falling masonry, as the wreck disposal teams were blasting the crumbled quays, but then several Z type mines broke surface and were rendered safe by the waiting US Mine Disposal Group.

Nearby, one BYMS and one MFV started the clearance of the Bassin Charles X. The BYMS was secured to the quay on the north-east side of the Avant Port, her LL tails were towed into the basin by motor dory, and anchored there. Then the MFV manoeuvred into the entrance, in order to operate her gear. They switched on, and two 'blue' magnetic mines detonated on the 22nd impulse of her LL tails, and coincided with the throwing of hand grenades into the basin.

Then clearance of the transatlantic dock went ahead fast. The MLs found some 'Katies' there, then BYMS 2155 entered and moored alongside a wreck. Her LL tails were towed into the dock and she slung her towed SA box over the quarter pointing south. On her second magnetic pulse, two 'red' mines detonated and, an hour later, when her SA gear was still running, but not her LL, an acoustic went up too.

The 'Katie' mines were causing much concern, and a special trawl sweep was brought over by trawlers from England. But no mines were found as the marine growth at the bottom of the undredged harbour was found to be incredibly thick. BYMS 2034 and 2213 reported reaping to be excellent, amounting to between four and six tons in one lift! A bottom sweep was also tried, using heavy cable, but again, wrecks and obstructions were frustrating to the sweepers.

Maintenance sweeping

So Cherbourg was opened up in short order, after this mammoth clearance operation, and very soon the harbour was crammed with Allied invasion shipping, more than had been seen in the harbour before the war or has been seen since. The famous 'red ball' truck route ran south from Cherbourg, and east to the front lines of the invading armies.

Routine sweeping went on continuously to ensure that all was well. In October, for example, the approach channels were being cleared each week, nine times by Oropesa, and 18 times with LL/SA sweeps; whilst the outer anchorage received six wire sweep passes and 11 LL/SA. The two inner anchorages were given 14 passes of each type of sweep. There was usually a BYMS or YMS flotilla there, plus a number of MMSs. About 15–20 mines a week were still being cleared, until the end of the year.

Le Havre

This was the second great port in the invasion area, lying on the east side of the Seine bay. But it never assumed the same importance to the invading armies as did Cherbourg, partly as it was liberated later, partly because Allied air raids had nearly demolished the port and the town in 1944, to bottle up or destroy German destroyers and E-boats which would have attacked the western flank of the invasion fleet.

A big clearance operation was needed, before it could be used by shipping, and October saw great activity there. The Thirty-first Canadian Flotilla was there, together with the BYMSs, YMSs, and MMSs. The same methods were used as at Cherbourg, and there were casualties too. The Canadian fleet sweeper *Mulgrave* was seriously damaged on October 8, and BYMS 2030 was sunk there the same day while carrying out LL/SA sweeping. The American YMSs, which visited Le Havre early in 1945, were lucky to get away with some near misses.

Clearing the coastal channel

First, the channel round the coast from Le Havre to Dieppe had to be swept clear, and the fleet sweepers concentrated on this. The Canadian Thirty-first Flotilla carried out the wire sweep, and the British Fortieth and Forty-second

Salamander, *of the British 'Halcyon' Class, after she and several consorts were sunk or damaged in error by RAF Typhoon fighters' rockets off the coast near Le Havre, August 27 1944.*

Flotillas followed close astern, doing the LL/SA clearance. The First Flotilla of 'Halcyon' was also active here, and it was whilst sweeping off the coast north of Le Havre that a tragic accident took place on August 27. A strike of RAF Typhoons, searching for enemy shipping, mistook them for Germans, and before recognition could be established, two ships, *Britomart* and *Hussar*, had been sunk and a third, *Salamander*, heavily damaged, as shown in our photograph.

Pressure sweeping by Stirling craft

It was thought that the Germans would have sown 'oysters' in the Le Havre approaches, in what was dubbed Operation Hoover. *Cyrus*, one of the two experimental Stirling craft, was used in clearance operations outside the breakwater. First, she was towed from the Thames to the Solent, where, on November 9, her operational tugs *Griper* and *Jaunty*, joined her, with the 'Catherine' Class fleet sweeper *Elfreda* in company, and in command.

The wind was blowing Force 5–6, and they had some difficulty in towing the barge out through the crowded anchorage. They started with both tugs alongside, one each side of *Cyrus*, then with one towing ahead and the other alongside. Outside the Solent, in the open sea, they towed with one ahead and the other astern, acting as a brake. They worked their way over to Le Havre without incident, and approached the special mooring buoy laid outside the harbour for *Cyrus*. But the remote control steering gear carried away at the crucial moment (as it often did), and *Cyrus* drifted down on to the American LST 295, and was impaled on her port bow. The tow parted, and the barge remained stuck fast on the port anchor of the LST for quite a while. In a hazardous operation in rough seas, and in the open roadstead, the tugs finally pulled her off using an 18-inch manilla hawser, and secured her to her buoy. The handling crew of nine aboard *Cyrus* were taken off amid sighs of relief. Sweeping was not always a battle of technology!

Her first operational sweep, on November 15 had to be abandoned, as her steering gear failed again. But the next day the approach channel was swept outwards and inwards in five hours, to a width of about 240 yards, but without result.

After a day's pause for heavy weather, another lap in and out was swept on the 18th, then two further days of winter gales followed. On the evening of the second day, *Cyrus* broke from her moorings, in a full westerly storm Force ten, with breaking seas 20 feet high. Desperate attempts were made by the two fleet tugs to bring her back under control; the waves were breaking right over the forward end of the barge, and the two tugs had their towing decks continuously

The Canadian 'Bangor' Class fleet sweeper Mulgrave *in Le Havre in October 1944, after being seriously damaged by a ground mine just outside the harbour.*

under water in the heavy seas. *Elfreda* pumped oil into the water around the barge, in efforts to damp down the big waves, but it was to no avail, and *Cyrus* eventually went aground in the mouth of the River Seine. Both tugs were damaged in going alongside the barge during these gallant attempts, and it seemed impossible to control this heavy barge in wintry weather.

Cyrus was refloated on November 29 and was then used, moored firmly, for static sweeping of the swept channel. A mile of buoyant LL cable was used in what must have been a miracle of seamanship in that weather. But four nights later, she broke adrift again in a full gale, and this time went aground and broke into two sections on the Banc d'Amfard, in the River Seine. This time salvage eventually proved to be impossible, and she was abandoned. So ended the saga of the two Stirling craft.

River clearance

Lastly, the sweepers needed to clear a channel all the way up to Rouen, well up the River Seine, which was navigable for large merchant ships and valuable for supply purposes. This was carried out with MLs leading the way up the wide river, followed by BYMSs, and the 131st Flotilla of MMSs. In some shallow areas, the river was cleared by MFVs, fitted with an LAA sweep.

Eastern Channel ports

A swept channel was cleared along the coast, as the ports were liberated and each small port was cleared as it was reached. Dieppe was cleared early in November, by a flotilla of ten MMSs, with little trouble. Boulogne, however, proved to be a different mix of mines. MLs and BYMSs tackled that one, and BYMS 2255 was sunk there on October 5, whilst at anchor. It was thought that a delayed-action moored mine must have been sitting right under her. BYMS 2154 was damaged there too, by a near miss on the previous day. Calais was cleared fairly simply, but while sweeping near there on December 6, some BYMSs were shelled from the vicinity of Dunkirk, the only port on this whole coast which remained in German hands until VE-Day.

Western Channel ports

In the summer and autumn, as the little ports along the north coast of Brittany were liberated, so they were cleared in turn by the sweepers. The same pattern was followed as up-channel; the MLs went first, with their shallow Oropesa sweep, followed by a BYMS or MMS flotilla, sweeping with both wire and LL/SA, until all was clear for shipping to start moving again.

All along this rugged coast there are sharp rocks above and below water, with one of the highest ranges of tide in the world, and sweeping tides to match. Under these conditions the need for accurate sweeping was greater than ever, but there was a great risk of parted sweeps and damaged LL tails.

Brest

Round the storm-swept north-west corner of Brittany lay this great French naval port, presenting another great minesweeping problem. The sweeping force was composed largely of the veterans who had started at the Normandy beaches—the Fourteenth Flotilla of fleet sweepers, the 159th of BYMSs, the 102nd of MMSs, half a squadron of American YMSs, the Fifth ML Flotilla, and the 713th LCPL flotilla, this last against snag mines. Added was a British minesweeper maintenance ship, *Sylvana*.

The tactics learned at Cherbourg stood them in good stead. The LCPLs led the way in, followed by the MLs with Oropesa, and the BYMSs astern of them. A mixed bag of magnetic and acoustic mines was cleared, several going off between the married portion of the LL tails. MMS 115 was damaged by a very close near miss, while YMS 231, testing her acoustic sweep late one night while moored to a buoy, put up a mine just 20 yards astern, which nearly sank her. Four other YMSs/BYMSs were shaken up by near misses, but they were all back in action within 24 hours.

The sweepers were operating under great strain here. As an example, the 102nd Flotilla of six short MMSs swept continuously for 114 operating days, without any of the ships suffering a breakdown—a great achievement. Only one sweeper was lost during the Brest clearance—LCPL 18, which sank under tow in heavy weather.

Clearance of the River Scheldt

This great river, winding up from the North Sea, leads to the port of Antwerp. Both the Allies and the Germans knew that once the port was captured by the armies, it would become vital to the supply of the advancing troops and armour as they liberated Holland and invaded Germany. Inevitably, a great effort was made by the Germans to render the river unusable and an equally great effort by the British to sweep it.

At one time, up to 112 minesweepers were working on the main river clearance, in addition to the 50 sweepers working outside on the approach channels. After the great assault on Normandy, this clearance must rank as one of the greatest minesweeping operations of the war, even on the same level as the major US operations in the Pacific, including the assault on Okinawa.

The Westerschelde, as the estuary of the river is called, is 60 kilometres long. At its mouth, between Flushing and Zeebrugge, it is up to five kilometres wide, but it narrows considerably on the way up. The channel is winding, with numerous sandbanks, the tidal range is considerable, and poor visibility especially in the winter months, is quite frequent. Altogether a setpiece for a major mine-

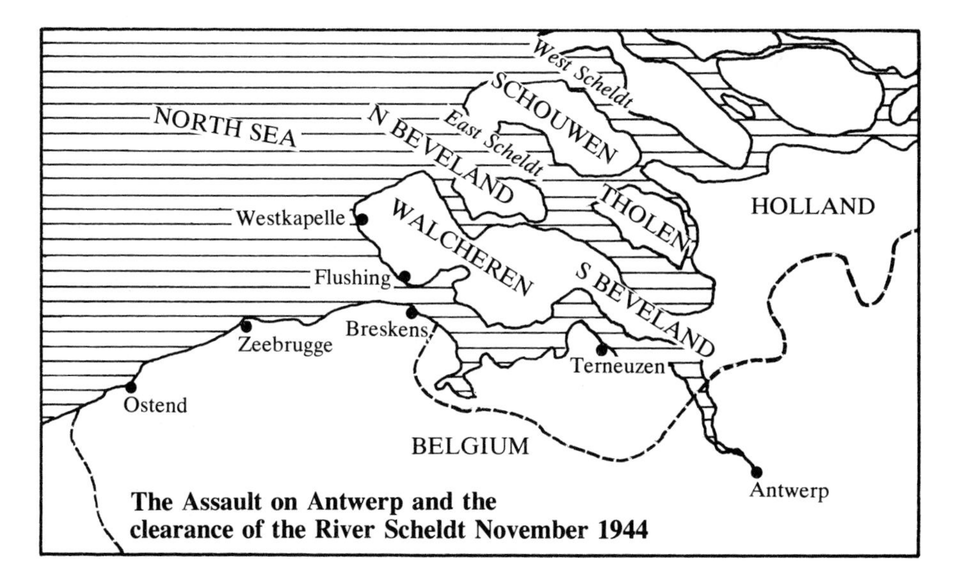

The Assault on Antwerp and the clearance of the River Scheldt November 1944

British MMS clearing ground mines inside the River Scheldt during the clearance of the way to Antwerp in November 1944.

sweeping problem, with large numbers of sweepers trying to operate with long tails in confined waters.

The plan was that, of the two major British minesweeping forces in the Thames Estuary, that from Harwich, called Force 'B', would be responsible for the sweeping of the outer approach channels to the river, while the Sheerness force, called Force 'A', would provide the sweepers for the river clearance. In the meantime, of course, all the Thames and East Coast war channels had to be kept swept as well.

An essential preliminary was the assault on the island of Walcheren, on the north side of the river entrance, near the port of Flushing. The landing was bitterly contested; but the sweepers swept in the bombarding battleship *Warspite* without loss. As the great ship passed inshore through the dawn mist past the tired sweepers, her signal light blinked: 'Your difficult task is done, my easy one is now beginning.'

On November 3 1944 Zeebrugge on the southern shore was captured. Force 'A' had left Sheerness on the 2nd, but on arrival, was turned back by shellfire from Knocke; three MMSs were damaged by shells, but the battery was captured on the following day, and the sweepers could proceed.

Then Force 'A' swept through the ships of Force 'B', who had completed their initial search of the approach channels, and swept on up river to Terneuzen, 12 kilometres upstream, which they were to establish as their base. They were to have been met by LCPLs from Ostend, with snag line sweeps, but the weather was too rough for them. Fifty mines were swept in the day and, in the meantime, the port parties had already arrived on the quays in Antwerp to clear them ready for handling merchant ships.

Over the following two weeks sweeping was carried out under full pressure. In the outer approaches the greatest problem turned out to be moored mines with heavy chain obstructors, which the MMSs could not lift. The Eighteenth Flotilla of fleet sweepers was sent in to drag these mines out of the channel.

There were great numbers of ground mines in the river, and they were

Above *Sweepers in harbour at Terneuzen, 12 miles up the river, during the clearance of the Scheldt in November 1944. LCPCL 278 is in the foreground, and among the ships behind are MMSs 11, 28, 146, and BYMS 2189.*

Below *A ground mine exploding in the River Scheldt, near Antwerp, in November 1944.*

detonating at quite a rate during these two weeks. The total bag in the river itself was 237, made up of 167 magnetic, 34 acoustic, and 36 moored. Up at Antwerp the port parties used portable magnetic pulsers and searched the lock entrances and narrow passages; they cleared 51 ground mines, as well as searching four million square yards of quay.

However, it was not a simple clearance operation. The E-boats, based only one hour's run by night up the coast at the Hook of Holland, re-mined the approach channels as fast as they were swept, and aircraft laid many more mines in the river by night. One hundred mine-watching posts, manned by 600 Dutch civilians, were put into operation on the river banks during the month.

The sweepers persisted. Dan-buoys laid at one mile intervals marked their triumphant progress, and double-flagged dans marked the positions where permanent buoys were to be laid. Their perseverance was rewarded. On November 26 the first three coasters arrived safely at Antwerp, but with everyone holding their breath. On the 28th, the river was declared free of mines and that same day 19 large merchant ships arrived in port. The great supply operation over the beaches in the Bay of the Seine had finally served its purpose; the sweepers had opened the way to a much shorter supply route.

Force 'B', which had swept 24 ground and one moored mine in the approach channels, returned to Harwich on November 28; but Force 'A' stayed in the river until December 23, before returning to Sheerness, sinking a midget submarine on the way for good measure. They lost only two ships in this great operation—ML 916 on a ground mine in the river, and MMS 257 on a moored mine in the outer channels.

One fascinating experiment was tried, when magnetic 'oysters' were suspected in the river. Two BYMSs, 2189 and 2188, anchored nine cables apart and streamed their LL tails. Then, with no movement to actuate the 'oyster', they pulsed their magnetic sweeps without weighing anchor. They swung on the ebb and the flood without interfering with each others' tails, and swept a different patch each time. BYMSs were used, as they were not dependent on batteries for LL sweeping, and the path swept was reckoned as 300 yards long, 600 yards wide, and magnetically swept to seven milligauss. The weather was atrocious, they needed to take shelter on several days, and it took 17 days to complete this sweep.

River Scheldt mine clearance force

Force 'A'—River clearance—from Sheerness.
1 Headquarters ship, with ML as tender
7 Trawlers, for stores, fuelling, and survey
1 MFV for LL sweep maintenance
3 Flotillas of BYMSs—30 ships
5 Flotillas of MMSs—36 ships, including 4 Dutch
2 Flotillas of MLs—16 ships, fitted for Oropesa
1 Flotilla of LCPLs—6 ships, fitted with snag line sweep
4 Flotillas of MFVs—24 ships, for river work
Total—112 ships
Force 'B'—Estuary clearance—from Harwich.
Fleet and motor minesweepers as necessary.
Total—50 ships
Total M/S force, at peak—162 ships.

MMS 32 sweeping off Baril, in the Mediterranean, in July 1944. Both her SA hammer and her LL sweep are out, and she is flying the appropriate warning signals.

Mediterranean assaults and clearance

On to Anzio

As the year turned, the assaults on Salerno, with both British and American sweepers engaged, settled down, and the sweeping fleet moved on towards the new assault on Anzio with clearance of the waters around Naples as an interim commitment.

The clearance of the Gulf of Gaeta was a long and arduous operation, but although sweeping conditions throughout were both difficult and dangerous, it was a complete success. The sweepers were operating close to the enemy-occupied coast and much of the sweeping was carried out at night, with the ships turning at the end of their laps, less than $1\frac{1}{2}$ miles from the shore. They were attacked by both shore guns and aircraft; ten ME 109s attacked the 'Algerine' *Cadmus* while she was weighing a dan but, although she was machine-gunned, the bombs missed her. *Fly,* another 'Algerine', had the second round of a shore salvo land only 50 yards ahead of her, although she was 20,000 yards from the nearest land. Bad weather grounded the fighter escort on a number of days, but a Ju 88 managed to shadow the sweepers, and four Fw 190s were shot down by escorting fighters during breaks in the weather. During the day sweeps the 'Hunt' Class destroyers *Hambledon* and *Lauderdale* steamed along inshore of the sweepers, in their support.

By this time, the sweepers had been operating continuously since the invasion of Sicily, and refits and breakdowns were taking their toll of the vessels available. So fleet sweepers were often acting as danlayers, and whalers were also extensively used for this duty.

MLs were much in the limelight, working ahead of the fleet sweepers. As at Normandy, a pair of MLs, with Oropesa sweeps out, went ahead of the Senior Officer's ship of each fleet sweeping flotilla, and other MLs followed the fleet sweepers, to dispose of floating mines. The Third ML Flotilla did great work here during this period.

Moored contact mines were in the majority in this area, and many had snag lines attached. The water was relatively deep and, at one stage, a shortage of deep-water dans threatened to slow down the clearance.

The Twelfth Flotilla of 'Algerines', led by *Acute,* and the Thirteenth Flotilla of 'Bangors', led by *Rothesay,* were the prime movers in this clearance. The names of the 'Algerines' raised memories of the days of Nelson in the Mediterranean—*Circe, Espiegle, Cadmus, Mutine, Fly, Albacore.*

A total of 157 moored contact mines was swept in a small area over a period of days; and it cost six Oropesa floats, seven otters, 58 cutters, and 25 dans! No sweepers were lost here.

Anzio

The next assault called for the same routine as before—a night sweep towards the beaches, swept channels illuminated by lighted dans, the bombarding ships passing through the sweepers as they moved inshore, and then the shallow water clearance of the boat lanes for the landing craft, by the smaller sweepers.

For the Americans, the sweeper group included the AMs *Pilot, Strive, Pioneer, Portent, Symbol, Dextrous, Sway,* and *Prevail,* while 14 YMSs and an SC provided the inshore clearance. For the British, the Twelfth and Thirteenth fleet sweeping Flotillas were there, joined by the Nineteenth Flotilla of 'Algerines', led by *Rinaldo.*

Not many mines were encountered, and the sweepers' main difficulty was operating in a crowded assault area. There was much gear fouled from near-collisions, and some escapes from floating mines, on one of which a YMS was lost.

During this period, the sweepers were keeping pace with the advancing armies on both the coasts of Italy. Many miles of inshore channel were cleared under shellfire so that supplies for the armies could be landed right up to the front lines. All types of minesweepers were in use for this work, but the MLs were again showing their versatility as sweepers. They started using double Oropesa sweeps in 'G' formation—another first, for it was at Malta earlier that MLs had first been used as sweepers at all. Ships' motor boats were also used in the clear water, to give the advancing sweepers warning of any shallow-laid mines. Much delay was caused by the anti-sweep explosive devices used by the Germans—in one spot, both the available MLs were out of action as sweepers for a while, as their gear had been parted by these explosive devices.

Assault on southern France

Operation 'Dragoon' was next, in August 1944. The assault itself was fairly smooth for the armies, but the large sweeping force did not have it so easy. Eighty-nine American and 50 British sweepers took part in 'Dragoon', and all of the 17 fleet sweepers allocated were among the British ships taking part. The

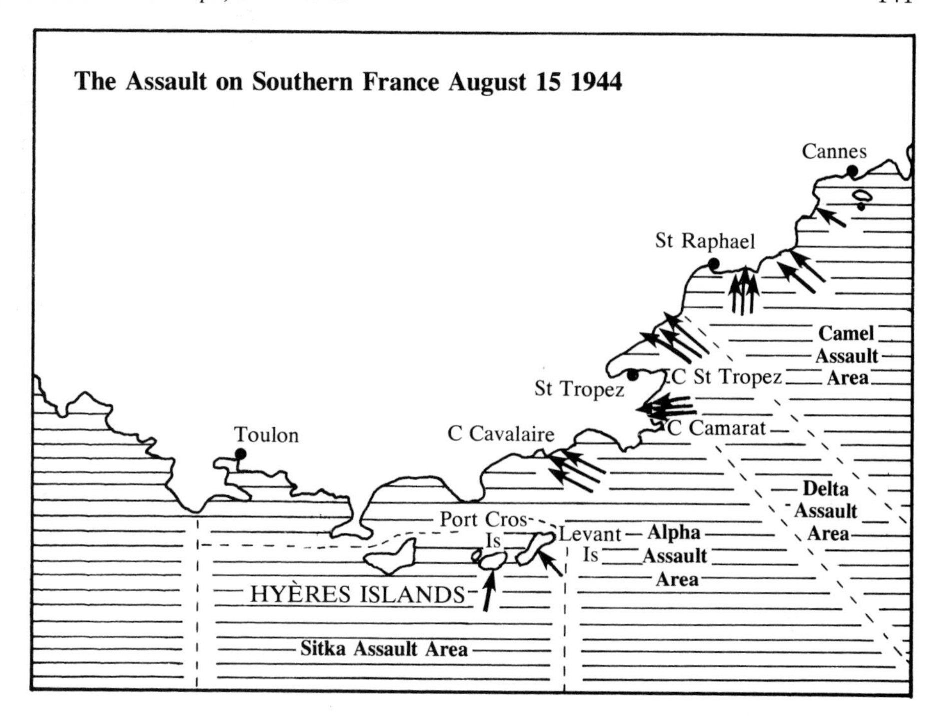

sweepers were initially divided between the four attack forces and, as soon as the initial assault was over, they were formed into a special task group to clear the coastal waters and the ports.

A dense field of mines was found during the assault on the Gulf of Frejus but, as the landings were progressing well elsewhere, this one was cancelled. The Gulf was cleared in three days, for the loss of four small sweepers. Over 200 mines were swept here. In 40 days, the entire South Coast of France had been cleared, including the important ports of Marseilles and Toulon. The clearance of Marseilles was a bit complicated, and took two weeks. 550 mines were cleared along the coast, mostly moored contact, except in the ports, and luckily there were no 'oysters'.

Greece
The clearance of Piraeus

The Germans had laid dense minefields to impede Allied progress in liberating Greece, and especially Athens. Three ex-Italian gunboats from Trieste were mainly used to lay the fields, until they were all destroyed. A large fleet of sweepers was engaged in this very dangerous clearance; two flotillas of 'Algerines', a number of BYMSs, and of GYMSs (BYMSs turned over to the Greek Navy), and a number of danlayers and MLs. In the operation, three GYMSs were sunk, together with ML 870, and two 'Algerines' were damaged, mostly in the initial clearance sweep of the channel into the Gulf of Athens. Hundreds of mines were cleared, again mostly of the moored contact variety. A little later, in the same area, BYMS 2077 was sunk, and all except seven of her crew lost, when she struck a moored mine on the edge of the channel being swept.

The British 'Algerine' Class fleet sweeper Regulus *sinking at speed, after hitting a mine during clearance operations in the Corfu channel in January 1945.*

The final German campaign

The Germans kept up their minelaying activities from Genoa and Spezia to the last, and this kept the sweepers very active. German minelaying craft made sorties from Genoa several times weekly, to replenish their minefields. Floating mines, carried on a current, were a favourite local tactic.

Both British and American sweepers were active in this area between the West Coast of Italy and the South Coast of France. The British sweepers were dubbed the 'Inshore Squadron', a name invoking memories of the days of sail. The Americans employed their Mine Division 32, led by *Implicit,* Mine Squadron 11, led by *Improve,* and Mine Division 18, led by *Sway,* totalling 12 AMs and nine YMSs, supported by the US Navy airship K 109 which could easily spot mines at five fathoms in the clear water (86 mines were disposed of by the 'blimp' calling up sweepers). There was continuous sweeping of this area right up to the end of the war.

Clearing the Adriatic

It was said at the time that, although not as glamorous as the big assaults, the clearance of this narrow sea was probably the biggest extended sweeping operation of the entire war.

From September 1 to December 5 over 2,000 mines were cleared in this area, and six flotillas of British fleet sweepers, totalling 39 ships, and large numbers of BYMSs, MMSs, and minesweeping MLs were employed. The strain on the sweepers was very great. The liberation operations ashore were long and difficult. The sweepers had to escort bombarding destroyers in, clear inshore channels for relief ships to get in to the civilian population, as well as in clearing identified minefields. They were continuously sweeping for up to 11 hours a day, with much loss of material—and some losses of sweepers also.

The final clearance in 1945

It is noteworthy that, although the fighting around the Mediterranean continued only sporadically during the first half of 1945, the minesweeping effort needed to continue at full blast. Only mines made convoys necessary in any part of the sea, and clearance was of both German and Allied mines.

The Germans did retain a foothold to the end of March in the northern Adriatic. On the 18th, three destroyers (one ex-Yugoslav, two ex-Italian) made a minelaying sortie; two British destroyers searched for them, and sank the two ex-Italian ships in the last destroyer action of them all in the Mediterranean.

As in the waters on the other side of Italy, mine-spotting in the clear water was effective and three flights of the old and slow Walrus amphibians were based in Italy for this work. More MLs were fitted out as sweepers, for shallow water work, and intensive sweeping was carried out with large numbers of mines being swept until all the waters were clear.

The final campaigns, 1945

The fighting in the southern part of the North Sea was by far the most intensive in that area of the entire war; the evacuation from Dunkirk much earlier had lasted but a few days, while this final battle ran for six full months, and was only ended by VE–Day. The German E-boats were based only an hour's run away at the Hook of Holland, the German small battle units worked from the islands north of the main river, and German minelaying aircraft worked over the area by night. The defending frigates and destroyers were based at Harwich, while the convoy escorts came from Sheerness, with MTBs based at Ostend and on the East Coast of England.

So the frigates, destroyers, and MTBs fought great battles with the German units by night, while by day a great fleet of minesweepers, based at Harwich and Sheerness, and with forward bases at Ostend and Terneuzen, worked ceaselessly to keep this vital ship channel open. The main NF channel (as it was called) had first been cleared after the capture of Ostend, by the Sixth, Seventh, and Fifteenth Flotillas of fleet sweepers. An extension, 20 miles long, was kept clear running north-east from the river mouth, for the fighting patrols to use at night.

Battles with minelaying E-boats took place on the majority of nights during the winter, in spite of the weather. The Germans suffered some casualties during these night fights, but more significantly, losses of Allied ships to mines went on throughout these months, despite the best efforts of the sweepers.

Towards the end of the year, the coastal forces control frigate *Duff* and an LST were badly damaged by ground mines, then the destroyer *Stevenstone*. The Forty-second Flotilla of 'Catherines' went to clear this field; with three ships sweeping LL/SA in line abreast, *Steadfast* detonated one mine just outside her LL tail, and *Combatant* two more, near NF 6 buoy. The fleet sweeper *Hydra* was mined off Ostend, and ten LSTs were sunk in this period, while the funnels and masts of sunken 'Liberty' ships dotted the area around NF 7 buoy. At Christmas time, the frigate *Dakins* was damaged, and the danlayer *Colsay,* anchored off Ostend, was sunk by a midget submarine.

Though only one-third of the year passed before the war ended in Europe, that was a continuing period of intensive minesweeping activity. While all the swept channels in the North Sea were kept clear, it was still that same channel over to Antwerp on which the main battle was centred.

The E-boats and midget submarines continued their minelaying sorties, though the fighting patrols were beginning to get the better of the battle by the spring. Until their fuel supplies ran out in mid-April, however, the E-boats continued their sorties on several nights a week. They continued to claim victims, too. By the end of April, 17 large ships had been sunk in this swept channel by mines, and three of the coastal forces control frigates, a third of the total of these vital ships, had been written off by ground mines, together with another destroyer.

In February, the E-boats laid one field containing 14 of their own mines, plus 21 British combination mines which had been recovered ashore. These latter had quite different characteristics from the German mines, and the British sweepers were lucky to escape damage from them.

In March, ML 466 was sunk by a mine while on patrol against midget submarines three miles north of the Westkapelle Light; and *Frolic* sustained much damage from three near misses soon after. Just as the war was ending, a small force of coastal sweepers, with the 'Algerine' Class fleet sweeper *Prompt* as headquarters ship, was sent over from England to clear the Dutch canals—but *Prompt* was mined on the way over, at the infamous NF 7 buoy, and had to be towed home.

The 'Egg Crate' displacement sweeps were brought into operational use at this time against 'oysters' laid by the German E-boats in the river approaches; the Fiftieth Flotilla consisting of 'Egg Crates' and their 'Bangor' Class tugs arrived at Ostend in March, and swept from NF 8 to NF 10 buoys in favourable weather. EC 10 operated regularly on this work, but EC 7 broke adrift from her tugs off Blankenberghe and was stranded.

The fleet sweepers were operating from Harwich and Sheerness at this time, since the anchorage off Ostend had been made unsafe by the midget submarines, and these were found also off Flushing, in the Downs, and off Margate.

A typical clearance operation was that carried out by the Forty-second Flotilla of 'Catherines' in January, after the frigate *Torrington* had engaged E-boats minelaying near NF 5 buoy. With *Fairy* (Senior Officer), *Combatant, Frolic,* and *Foam,* and the 'Isles' Class *Hermetray* as danlayer, they first carried out a high percentage wire search and cut nine mines, then followed up with an LL/SA search and lifted five more.

Up to five flotillas of fleet sweepers with numerous BYMSs and MMSs were operating in this crowded area in these four months, due to the importance of this one channel. They swept 122 ground mines and 21 moored contact, while the river sweepers based at Terneuzen got a further 90 ground mines and 16 moored.

Anti-submarine sweepers

The midget submarines were very active in this area and, although they suffered heavy losses from the patrols and the bad weather, they persisted. Some carried mines (one went ashore in the river entrance on March 7, and proved the point), but the sweepers kept a watchful eye open, and got their own back. Early in the year, for example, three BYMSs, 2213, 2141, and 2221, sweeping off Ostend, sighted several midgets together; they had no Asdic gear and no depth charges and were sweeping LL/SA in 'Q' formation, when 2221 reported a midget to port, and 2213 saw two more astern. They all steered to ram and opened fire with their Oerlikons; 2141 took a turn round one surfaced midget with her LL tail, caught the submarine's conning tower in the bight and hauled her in under the stern—but the midget foundered. Another midget was sighted one cable away to starboard, and zig-zagging; 2213 steered to ram, but the conning tower hatch opened and one crewman climbed out waving, just as the sweeper rammed. The midget disappeared under the bow, came up on the other side and then sank. The crewman was left clinging to the sweeper's LL tail. On another occasion, BYMS 2213 rammed a midget and sank it, while 2141 managed to tow one into port.

Keeping the Scheldt clear

Aircraft minelaying by night continued right to the end of the war up the river, and the sweepers based at Terneuzen were kept constantly busy. Many sweeping days were lost in the wintry weather, but 36 mines were swept in January. On the 30th of the month MMS 248 was sunk by a coarse acoustic in the river, 30 seconds after her hand grenade sequence had finished. The explosive sweep was now accounting for one in four of all acoustic mines detonated, and was effective in working off the pulse delay mechanisms.

Forty-four mines were swept in February, of which 35 were acoustic. There was much concern about the possibility of the unsweepable 'oysters' blocking the river, and the most intensive operational displacement sweeping of the war took place at this time. Apart from the 'Egg Crates' operating out of Ostend, the First Steam Gun Boat Flotilla was also based there, carrying out high speed magnetic-displacement sweeping. Ostend was being developed as a major minesweeping base, and *Sylvana* was there as maintenance ship, with ten MMSs, nine BYMSs, and many MLs.

The MMSs found much difficulty in clearing the water around merchant ships in shallow water. The navigational buoys were numerous and the tides fast, and the long LL tails kept on getting caught up. The MLs were by now carrying out all the wire sweeping inside the river, while the BYMSs and MMSs concentrated on LL and SA. In the winter weather, only the BYMS could operate easily outside the river.

So the great minesweeping effort continued without let-up right up to VE–Day, and then the great clearance operations of the postwar period started—the only difference for the sweepers was that the minelayers had gone, but the danger from the mines was no less for many months.

Chapter 7

The Pacific, 1944–1945

The great American island-hopping assaults of 1944 and 1945, starting from the outermost Japanese-held groups of islands and moving inwards inexorably towards the mainland of Japan itself, form a very different minesweeping story to that of northern Europe and the Mediterranean. There, the distances were relatively small and, if damaged sweepers could be brought back to port, there were repair facilities readily available and sweep gear could easily be replenished. But in the Pacific campaign, distances from a main fleet base to the landing areas were usually to be measured in thousands of miles, presenting enormous problems of logistics and repair. It was just as vital, before an assault, that the sweepers did their job thoroughly and on schedule, but it was much more difficult to accomplish. There was a further complication not encountered in Europe—the great campaign of the Kamikaze suicide aircraft, culminating in the assault on Okinawa, where warship losses from this one cause rose to remarkable figures. The Royal Navy's overall losses in sweepers were higher than those of the US Navy, mainly because they had been at war longer; but the Americans' rate of loss, especially in the waters around Borneo and Okinawa, was indeed high for a relatively short period.

It is interesting that the numbers of minesweepers employed in the main American Pacific assaults were lower than in similar operations in Europe. The total sweeper fleet at Okinawa was 116 ships, and at Balikpapan 46, whereas those used in the Normandy assault totalled 307, and in the clearance of the River Scheldt 170. The great American shipbuilding programme churned out excellent new sweepers at a very fast rate between 1943 and 1945, but the appetite of the Pacific island assaults—some running in parallel at the same time—was equally great. There was, of course, a continuous stream of new sweepers commissioning, being shaken down, and on passage to the far-flung Pacific battle areas, but the European war still absorbed quite a number up to May 1945.

Even in a specialised history of this size, space does not allow the story of the sweepers at all the island assaults to be told. So, to give the reader the flavour of the whole within this one chapter, it is necessary to concentrate on three aspects:

First, the Borneo operations—Tarakan, Brunei Bay and Balikpapan—as representative of typical large island assaults, but without the Kamikaze diversions of Okinawa.

Secondly, a summary of the other island assaults, showing roughly the numbers of sweepers engaged in each. The US Pacific Fleet was so large, flexible and far-flung, that no daily state of sweeper dispositions seems to have existed (the historian is fortunate that the Royal Navy did keep one), and from the distant standpoint of Europe, it has not been easy to piece all the details together, although copies of the US Navy's action plans and reports for the major assaults are in the records available in London.

Thirdly, an account of the sweepers at the assault on the island of Okinawa—the bloodiest attack of them all, during which actual minesweeping took second place to survival against the onslaught of the suicide planes. This action, as far as the sweepers are concerned, ranks alongside the invasion of Normandy for sheer size, action and gallantry.

The Borneo assaults, May–July 1945

Tarakan, April 22–May 3

The sweeping side of this assault was interesting, not for the number of mines swept, nor for the number of sweepers engaged, but because it was a very difficult sweeping problem from a technical point of view. The currents were exceptionally strong, and the YMSs could make little headway over the ground against them. This meant that moored mines cut by the wire sweeps became caught in the sweep gear, where normally the sweeping speed would have

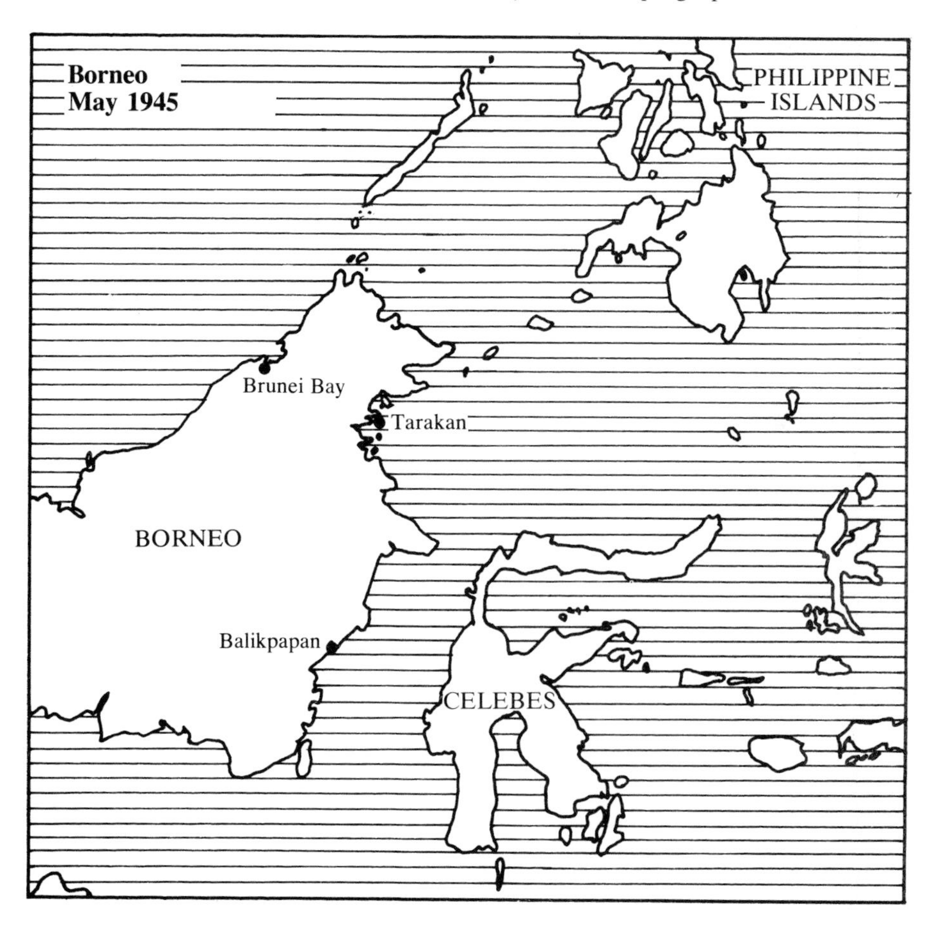

broken them free. However, at slow speeds the mines were often dragged willy-nilly into water which had already been swept.

Another problem was the variety of mines to be swept. The Japanese mines in the area were of the moored contact variety; but many American influence mines, both magnetic and acoustic, had been laid by air over the previous two years, and were indeed difficult to sweep. The same problem was also encountered at Balikpapan. Ship-counts up to seven 'clicks' had to be worked off before an area could be declared free of mines.

The currents added to the problem of sweeping the acoustic mines, as it was feared that the water movement might have added up to four feet of silt over the mines since they were laid, so they would have their 'ears' insulated from the acoustic sweeps. If this was the case, it was possible that later the currents could move the silt again, and live mines would be exposed in the swept area.

The Minesweeping Unit, TG 78.1.5, was part of the Tarakan Attack Group, TG 78.1. The sweepers numbered just 16—11 YMSs and one APD. This was a new destroyer escort, converted to a fast troop transport, and carrying in special davits four LCVPs fitted for light rope sweeping against moored mines.

The M/S Unit sailed from Leyte on April 22 1945, and arrived off its objective on April 27. It included a Hydrographic Unit, since the channels needed careful surveying and buoying once they had been swept.

The pre-assault sweeping started immediately, with the sweepers covered against shore gunfire by a Cruiser Covering Group, right up to the moment of the landings four days later.

The first sweep was an exploratory one, to clear an approach channel 30 miles long, from the 100-fathom curve to Point Whirlaway, near Tarakan. No mines were cut or lifted on this initial sweep. Later on the same day, three of the four landing areas were swept but no mines found, although four magnetics were lifted near by. The little LCVPs, with their rope sweeps, also cut two newly laid Japanese moored contact mines, right inshore. That night, following normal assault practice, the sweeping force retired out to the safety of clear sea, returning before dawn next morning.

Next day, April 28, three more magnetics were lifted, and seven moored contact mines were cut. Two more magnetics were known to be there, but did not respond to the sweeps. Another magnetic exploded only 30 yards astern of YMS 329, badly damaging her.

April 29 was P–2, or two days before the landings. It was lucky that the sweepers were not being fired on from ashore, because the swept channel was not considered sufficiently clear for the passage inshore of the Cruiser Covering Force, and it had to stay far out at sea.

But the sweepers were making progress; two of the three landing areas were re-swept, with no mines found, although nine moored contact mines and one floating mine were disposed of just four miles east of the area. A check sweep, extending 1,000 yards on each side of the approach track, also produced a clean sheet. In one landing area the YMSs went in to within 1,200 yards of the beaches themselves, while the little LCVPs dared to approach within 400 yards of the pier on two occasions, and escaped unscathed—but with no mines found either. However, there were still seven known magnetics in the area which were not responding to treatment.

April 30 was the last day before the assault, but it was decided that the preferred approach track was still not safe for the main force, so an alternative track which had also been swept would be used.

That day, magnetic and acoustic sweeps were continued under great pressure in all landing areas, while destroyers closed in to give fire support to the sweepers, as the Japanese were expected to start reacting to such a daring approach. During these operations, the destroyer *Jenkins*, while retiring from a beach, hit a Japanese moored contact mine 1,000 yards off the centre of one approach track. Her casualties fortunately were light, but she was extensively damaged and lay dead in the water until she could be towed away. Two more contact mines were cleared in the same spot soon afterwards, and two YMSs received minor damage from other exploding mines.

That night, the Attack Group passed safely through the swept area, on its way to the beaches—and four influence mines were swept in their wake, just after they had passed!

Then, at dawn on May 1, the landings went in. For the sweepers, it was a surprisingly quiet day, and that night the APD, *Cofer*, and the YMSs reported for duty in the defensive screen thrown around the beaches; a close watch was kept for Japanese small craft and swimmers.

May 2 saw the scene change again, and the sweepers were in trouble. Eight YMSs were making a combined check sweep through a narrow strait, three in the van using Oropesa wire sweeps and five astern with their magnetic sweeps out. Japanese shore guns, probably 3-inch, and in well-camouflaged positions, opened fire on them in the afternoon, and YMS 481 (the very last ship of the class to be built) received several direct hits. One exploded on the stern, and the magazines went up, and she sank. YMSs 334 and 364 also received hits while retiring hastily out to sea, and the APD and two support landing craft moved inshore and silenced the batteries. YMS 364 was straddled and had numerous shell holes in her hull and superstructure, while YMS 334 had extensive structural damage.

But next day, May 3, the landings succeeded, and the sweepers were able to move on. An LST group left that day, after landing their troops and tanks, and two of the YMSs were in their sea escort. During this short period of intensive sweeping, 43 mines had been accounted for, at the loss of one YMS sunk and a further five YMSs damaged from a minesweeping force of 11 ships.

The Commander of the Amphibious Force (TF 78) commended the sweepers for their cool conduct under fire, commenting that the performance of the Minesweeping Unit was uniformly excellent and had made a material contribution to the success of the operation.

Brunei Bay, June 7–16

So the assault force moved on, just one month later, with Australian troops once again making up the landing force. This assault also posed a difficult sweeping problem as large numbers of Japanese mines had been planted deep, in a tanker anchorage, to keep out Allied submarines. The number of mines swept was far higher than at Tarakan, and this time fleet sweepers (AMs) were also taking part.

The Minesweeping Unit was designated TG 78.1.5, with MinDiv 34, composed of five AMs—*Sentry, Salute, Scout, Scrimmage* and *Scuffle*—and the YMS Unit— *Cofer*, the APD with her four LCVPs, 12 YMSs and one tender, an LSM, which carried in her hold two 50-foot motor launches, fitted for sweeping right inshore in shallow waters. So there were 23 sweepers in the force, plus one further YMS, 160, in the Hydrographic Unit.

The sweepers were disposed like this: TG 78.1.51 (Sweep Unit 1)—five AMs; TG 78.1.52 (Sweep Unit 2)—four YMSs; TG 78.1.53 (Sweep Unit 3)—four YMSs; TG 78.1.54 (Sweep Unit 4)—four YMSs; and TG 78.1.55 (Sweep Unit 5)—one APD and one LCM.

The Hydrographic Unit was led by the Australian frigate *Lachlan.*

The M/S Group sailed from Morotai on June 2, and arrived at its objective early on June 7—again showing the distances involved.

That day, the AMs began their sweeping operations, under the protection of the Cruiser Covering Force and Fire Support Group; 34 moored Japanese mines were cut, as they commenced clearing the swept channel miles out from Brunei Bay itself. Most of these mines had been planted in a line at the south-west entrance to the bay, and four of the landing areas were swept without result.

Next day, June 8, the sweeping continued on schedule, and 34 more moored contact mines were cut in the same area, by the AMs, although the YMSs, moving inside the bay, found none. In the afternoon, the AM *Salute* struck a moored mine in water already swept and was seriously damaged, suffering six killed, three missing and 37 wounded. She sank later that night.

June 9 was the crucial day before the landings. Six separate areas received check sweeps, with no mines found, although a few were cut on the second time round. The AM *Scuffle* reported mines exploding in her gear when streamed to a depth of 60 feet, perhaps indicating a new type of Japanese anti-sweep device.

The frigate *Lachlan* went right inshore to make a reconnaissance of one beach, but found no mines.

The landings took place on schedule, at dawn on June 10, and no mines were swept that day, nor ships lost to them. But the routine sweeping continued and next day the YMSs cut 25 moored mines in just one area, although four other regions had produced a clean bill. The sweeping continued to be difficult in technical terms—nine sets of wire sweep gear were lost that day on uncharted rock pinnacles. Support landing craft (LCS(L)s) were now being sent in to sink the cut mines by gunfire; four, plus two destroyers, *Philip* and *McCalla*, were engaged in this work, especially a few days later in the big field at Miri.

On June 12 five more mines were cut, to break the 100 mark in mines swept since June 7.

The deep anti-submarine minefield was tackled next, starting on June 13. First the AMs went in, with their Oropesa wire sweeps set deep, and cut 31 moored mines on their first leg, although five sweep wires were cut by mines exploding in the gear. In sweeping later that day they cut a further 61 mines, making a fine bag of 92 for the day. 3-inch shore batteries opened up on them from time to time, 25 rounds being fired although all fell short. The destroyers opened up with counter-battery fire, although they could not edge in too close due to danger from the floating mines cut by the sweepers.

Twenty-nine more were cut on the first leg of the morning next day, June 14, and the bag for the day was a magnificent 164 mines. But this deep field was wreaking havoc on the sweep gear. *Scrimmage* parted all of hers, while *Scout* and *Sentry* were left with their starboard sweeps only. The YMSs were experiencing such difficulty that they had to attach explosive cutters to their sweeps (a practice which already had become largely standard in Europe), and when these mines became caught in their gear, the ships were slowed down almost to a stop. The mines in this field had been deeply set and thickly planted, and posed a really difficult sweeping problem.

By June 15, the mines cleared in this one area near Miri added up to a surprising 246, with the field laid on a continuous line from shoal water to nine miles offshore. The AMs had suffered such heavy losses of their wire sweep gear that they were recalled to Brunei Bay, and the YMSs carried on with the sweep. Many sets of their gear had been fouled, and the mines had heavy chain moorings. However, it was decided that, although losses in sweep gear were enormous, it was not advisable to stop to attempt to recover parted sweeps. Instead the line of sweepers was to re-arrange itself and do the best possible job with those sweepers which still had operative gear. Nevertheless, by June 16 the field had largely been cleared, for a bag of 440 moored contact mines, 338 swept off Miri and 102 in Brunei Bay.

Balikpapan, June 15–July 6

This was a truly interesting operation, including in the large force of minesweepers three AMs, 39 YMSs, one APD and one LCM. First, the approach was long, shallow, and heavily mined, needing a complicated minesweeping operation for 16 days before the landings, so there could be no element of surprise. Secondly, the mines laid presented a hazardous mixture—Japanese moored and ground influence, and Allied influence. The latter presented a special problem: they were very sensitive magnetics and acoustics, set for up to seven ship-counts, and had been planted in water with an average depth of 3½ fathoms. They had been laid some nine months earlier, by Australian bombers, and totalled 54 British and 64 American mines, none of which had arrangements for sterilisation. It was realised later that to lay unsterilised mines in an area which would afterwards be assaulted created a great additional hazard. In addition to these mines, there were Japanese mines laid at the harbour entrance, in a line 6,000 yards long, and these, too, presented a special problem as there were many shoals and the enemy shore defences were very strong.

The timetable for the pre-assault sweeping was pretty tight. In that shallow water, even the well-degaussed wooden YMSs were in great danger, and their gear was constantly at risk due to the uneven bottom and the numerous wrecks.

The best magnetic sweep available could in those conditions only sweep a path 100 yards wide, and so the sweepers had to navigate with great accuracy if extensive 'holidays' were to be avoided. The shore was poorly marked for navigation, and some of the fleet destroyers did great work in moving inshore and assisting the sweepers to be sure of their navigation, using their better radar and navigational aids.

Yet another special hazard was the persistent and accurate Japanese shore gunfire, which constantly disrupted sweeping operations. The YMSs were using a catenary magnetic sweep, one towed in a loop between the sterns of two or more ships, and in retiring at speed when under shellfire they found it necessary in many cases to jettison the magnetic sweeps. This gave rise to a critical shortage of magnetic sweep cables.

In the early stages of the operation, when the shore gunfire was at its most accurate, the Cruiser Covering Force had to remain some 12,000 to 14,000 yards offshore, due to the shallow mined waters, and it was difficult for them to neutralise these shore guns effectively at that range.

The original force of sweepers allocated in the planning stage—four AMs and

16 YMSs—was soon found to be inadequate. Further YMSs fitted with magnetic sweeps were rushed up, although some could not be released from Brunei Bay until that operation had been completed.

The minesweeping plan divided the known minefields into four sections: **1** all waters outside the 100-fathom curve, where no mines had been reported; **2** a small area where the presence of enemy shore-controlled mines was suspected; **3** a field of single ship-count Allied mines; and **4** multiple-count Allied influence mines. It was felt that Japanese mines, moored or influence, could be mopped up relatively easily, with the Allied influence mines presenting the major problem.

The exploratory sweep would cover a path five miles wide from the 100-fathom curve in to the beaches, to cover moored and acoustic mines, and afterwards the magnetics. The shallow water, extending a long way offshore, meant that the little YMSs had to cover all the sea area inside the ten-fathom curve, since the motor launches and LCVPs carried respectively in the LCM and the APD were not of use here, due to their slow speed and the fast currents. The AMs, coming in after the first YMS clearance, were to sweep all the water outside the ten-fathom curve.

Casualties among the sweepers in the assault period were high. Three YMSs were sunk and five damaged. The three sunk were all caught by mines, but shore gunfire accounted for three of the damaged ships.

Another interesting aspect of this operation was the lack of suitable minesweeper store and maintenance ships carrying adequate replacement sweeps. This had a serious effect on the success of the operation (which was 460 miles from the nearest supply base). It is remarkable that this should have happened a full year after the Normandy landings, where both the British and American navies had correctly foreseen the need for these support ships. The British Pacific and the East Indies Fleet were provided with specially converted ships for this purpose, with large stocks of replacement sweep gear of all three types, as well as of LL generators. The United States Navy after the Balikpapan operation recommended the use of LSTs in the Pacific for this specialised duty.

The M/S Unit was designated TF 78.2.9 in Task Force 78, and consisted of three AMs (*Sentry, Scout*, and *Scuffle*), 39 YMSs, one APD (the faithful *Cofer*), and one LCM. The Hydrographic Unit sailing in company included the Australian sloop *Warrego*, and YMS 196, the latter experienced in this role from the Brunei Bay operation.

The first element of the unit, 16 YMSs led by YMS 392, left Morotai on June 11 for Balikpapan, with a destroyer as escort and *Cofer* with her four LCVPs embarked.

They arrived offshore from their objective on June 15 and started on the approach channel sweep from the 100-fathom curve, sweeping down to 60 feet for moored mines, with their acoustic sweeps out as well. No mines were found as they moved inshore to the ten-fathom curve, the magic line beyond which trouble lay. The exploratory sweep continued next day, with similiar negative results, while the LCVPs swept in one area to six fathoms—an encouraging start, although two YMSs had already had to withdraw for repairs.

On June 18 the sweepers moved into dangerous water—and in more ways than one, for as well as finding mines they were fired upon by the shore batteries. YMS 50 lifted a magnetic mine right beneath her, and nearly broke in half; but the shore batteries opened fire to prevent the LCVPs from towing her out, so they had to content themselves with rescuing the survivors while the wreck was sunk by gunfire.

The YMSs were now finding it necessary to jettison their loop magnetic sweeps when under heavy shellfire; four sweeps were lost and two damaged on that day alone. Next day, June 19, they were again dispersed by heavy shelling, and a further four magnetic cables had to be jettisoned.

YMS 368 lifted an Allied magnetic mine too close alongside on June 20, lost her sweep, and had her gyro compass thrown out. She had to retire. Next day the sweepers were fired upon three times, and sweeping was disrupted twice. YMS 335 was hit twice on her fo'c'sle by shells, her 3-inch gun being seriously damaged, with four dead and six wounded. In the covering fire to silence the shore batteries, supporting warships, still far offshore, fired 1,347 rounds of 6-inch, 270 rounds of 5.9-inch and 451 of 5-inch ammunition. YMS 392 recorded that 25 to 40 shells from the shore fell close to the ship, some as close as five yards. She slipped her magnetic sweep, increased to full speed, and 'resorted to radical measures'. One shell landed right underneath her fantail, and in rage she fired several rounds in return from her own 3-inch, all being seen to land on the beach.

Shortage of magnetic cables was by now becoming a critical problem. Only six YMSs were serviceable on June 21, and it was difficult to plant channel marker buoys accurately in their wake due to the many sharp turns which they had to make under fire.

Reinforcement sweepers, both AMs and YMSs, were arriving by this time, and despite the fact that they brought some replacement sweeps with them the situation became increasingly critical. On June 22 four more cables were lost, and on the 23rd only an average of three magnetic sweepers could be put into operation at any one time. The shore firing continued each day: YMS 10 was holed above the waterline on June 22, and next day YMS 364 received a 3-inch shell hit directly into her bridge structure. The shell actually landed in the Captain's bunk, but it was smartly thrown overboard before it exploded! The strain, however, was telling a bit; 364 needed to transfer four men out as 'battle fatigue' casualties, and more cases followed from the sweepers, showing the strain this fierce battle was causing among the gallant crews.

An additional hazard materialised in the evening of June 25, when four Japanese aircraft attacked the sweepers with torpedoes during their first night inshore at anchor; three of these attackers were shot down promptly and the fourth disappeared—*Cofer, Sentry* and a YMS claimed the honours.

But the sweeping was producing results, however difficult it was, for it was vital that the water be cleared before the assault went in on July 1. By June 26 the five-mile-wide approach channel had been proved clear, and at last the supporting warships could move in closer to silence the shore batteries so that the minesweepers could get on with their work.

That same day, YMS 365 exploded a magnetic mine close to her sweep cable, then another mine lifted right underneath her and two other YMSs had to take off her survivors just before she sank. Then YMS 39 lifted one right under her keel and sank in less than a minute, suffering three killed and seven wounded. Loss of magnetic cables was still a grave problem, although some of those jettisoned under fire had been recovered and repaired. LCM 1 had a minesweeping repair unit embarked, and did sterling work. But she had no spare cables on board for her hold was largely taken up with the two big 50-foot motor launches, and even had she carried spare cables, she had no equipment for transferring them to the YMSs. Some spare cables were, however, rushed in from the fleet bases, by destroyer, APD and AMs.

Only four days now remained before the assault, and it was decided that, of the four main landing areas, two could be cleared with up to five ship-counts by F-day (July 1), but the other two could not be adequately covered against the magnetics. On June 28 two more YMSs retired from the scene, 72, one of five reinforcements en route, by running aground, and 49 by lifting two magnetics very close, flooding her stern compartment.

Enormous efforts were being made to keep to the assault sweeping schedule, and sweepers were operating repeatedly in unswept waters. By now, of the 15 YMSs still there, nine were able to operate magnetic sweeps. Three others without magnetic sweep cables were being used as Oropesa wire sweepers, and three more were under repair, so the picture was improving a little.

By June 29 the assault approach channel had been swept to the 15-foot mark, and was considered to be cleared. YMS 10 exploded a moored mine in her wire gear, then found two more in the sweep when recovering it. But although two more magnetic cables were damaged that day it was a good time, for five replacement cables arrived.

By the evening of June 30, the eve of the assault, 27 mines had been cleared. Note how difficult it was for the YMSs to sweep the sensitive Allied magnetic mines with their loop sweeps. Comparing this relatively small bag with the list of ships sunk or damaged, it can be seen that one YMS came into that category for every three Allied mines accounted for. Even at this point, there were some areas in the landing zone which had to be declared 'no go' for both the assault and the covering force as a result of unswept mines.

Then came the landings, which were right on schedule and completely successful. But the sweeping continued without a pause. On July 2 YMS 47 was damaged and had to pump like mad to keep the incoming water under control until a rescue ship came to assist and tow her away. The sweep unit commander reported this day that he had 24 YMSs, of which six were leaving the area, five were operating magnetic sweeps, four had just arrived and were testing their sweeps, and nine were undergoing repairs to their magnetic sweep gear.

But the clearance was continuing apace as the number of operative sweepers rose (ten YMSs were operating as magnetic sweepers on July 3), especially now that they could work more safely without enemy interference. But even after the enemy had surrendered in this area, the sweeping went on, and more mines remained to be cleared. On July 9 YMS 84 was sunk by a Japanese moored contact mine off the port—the last sweeper casualty among these hotly disputed islands.

On July 6 the sweepers triumphantly cleared a channel into the harbour itself, cutting 18 Japanese moored mines without damage to themselves.

To summarise this operation, 34 moored contact and 16 ground influence mines constituted the total bag. Three YMSs were sunk and one damaged by influence mines, while three more were damaged by shore gunfire. Even allowing for repaired magnetic cables, 15 sets of this gear were lost off Balikpapan.

A sequence of Pacific island assaults

These were the more important of the United States Navy's series of assaults on Pacific islands. There were other minor landings in addition.

Guadalcanal—August 7 1942.
Bougainville—November 1 1943.
Gilberts and Marshalls: Makin—November 20 1943; Tarawa—November 20 1943; Kwajalein—January 31 1944; Majuro—January 31 1944; Ebeye—February 3 1944; Eniwetok—February 19 1944.
New Guinea: Lae—September 4 1943; Port Moresby—December 6 1943.
Admiralty Islands: Los Negros—February 29 1944.
Hollandia: April 22 1944; Morotai—September 15 1944; Peleliu—September 15 1944.
Marianas: Saipan—June 15 1944; Tinian—July 24 1944; Guam—July 21 1944; Ulithi—September 25 1944; Iwo Jima—February 16 1945.
Philippines: Leyte—October 15 1944; Ormoc Bay—December 7 1944; Mindoro—December 15 1944; Lingayen Gulf—January 2 1945; Subic Bay—January 29 1945; Corregidor—February 13 1945; Palowan—February 28 1945; Zamboango—March 8 1945; Negros—March 18 1945.
Borneo: Tarakan—May 1 1945; Brunei Bay—June 8 1945; Balikpapan—July 1 1945.
Okinawa: April 1 1945.

The island-hopping assaults

This was one of the most remarkable, and indeed heroic, naval campaigns in history. But it would be impossible, within the span of this book, to describe all the individual assault actions in any detail. They have been splendidly described in the United States Navy's official history of the period; here, it is only possible to highlight the minesweeping side of a few.

The picture which emerges is one of early struggle on the sweepers' part, for there was no ready made force available, other than the gallant little band of converted World War 1 'four-stacker' destroyers which performed magnificent service in the early landings, some going right through to Tokyo Bay. The assault forces had to wait for the trickle of new sweepers, both the larger DMSs and AMs and the small YMSs, from the shipyards in the United States to become the stream which it later became.

The other striking feature of all these operations was the enormous distances involved across the Pacific Ocean. From the fleet's traditional main base at Pearl Harbor, the sea miles were measured in thousands to the captured forward bases, such as Ulithi and Leyte, and still further to the landing beaches. It was usually up to a week's voyage from the landing force's forward base to the beaches, and the logistic problems involved in supplying the sweepers in such a situation are clear from the Borneo landings.

The campaign starts, 1943

Since the assault programme ran from late in 1943 through to mid-1945, it is convenient to start the island-hopping story with those landings which took place towards the end of that first year. New Guinea was the first area to be re-taken after Guadalcanal.

The assault on Lae was launched on September 4 1943, and eight days later the area was in the hands of the Australian troops. Five of the new YMSs were present—YMS 11, 49, 50, 51 and 70, and as there were no mines to be found they acted as escorts as well as sweepers.

Port Moresby, on the same island, was next, early in December, and here an

Australian fleet sweeper, *Katoomba*, led a unit of four further YMS—8, 46, 47 and 48—which met a hazard that was to be all too familiar in the later assaults—Allied mines. Here, they cleared an Australian-laid field of some 250 moored contact mines.

The Admiralty Islands came next. In February 1944, a small landing operation was successful.

It was supported by two of the old destroyer sweepers, plus five YMSs, which cleared the harbour without trouble.

Hollandia followed, a large group of ships making three simultaneous landings, and once again the sweepers had an easy time. Four destroyer sweepers and four YMSs were in that force.

The Gilberts and Marshalls came next, and at once we find island names which are now famous in American history. The landings started in 1943, when both Makin and Tarawa were assaulted on November 20. Makin was another easy one, with just a small landing force. It was swept in by the new fleet sweeper *Revenge*, a symbolic name which was to be in the van of the landing forces all the way to the Japanese surrender. No mines were found, however, either in the approach or in the big lagoon itself.

Tarawa was a different matter—the Japanese offered some fierce resistance. Two of the fleet sweepers, *Pursuit* and *Requisite*, swept the force in, *Pursuit* running an Oropesa wire sweep first, *Requisite* following astern with her magnetic and acoustic sweeps streamed. Although it turned out that there were no mines, there was plenty of gunfire from the shore, and smoke screens were laid around the sweepers by small craft, while the ships replied with their 3-inch guns. The weather was a bit hazy, which helped, but an aircraft had to be used to guide *Pursuit* to the shoals so that she could lay marker buoys for the following landing craft. Then, when the water proved clear of mines, they served as marker and control ships for the landing craft, tugs when required, and maids of all work.

Next came twin assaults on January 31 1944, on the islands of Kwajalein and Maduro. The latter was launched first, and the sweep unit, including the fleet sweepers *Oracle* and *Sage* and two YMSs, found no mines. Nor was there any resistance ashore.

The M/S Group for Kwajalein was somewhat larger—five destroyer sweepers, three AMs and four YMSs. Four of the destroyers were with the Main Attack Detachment (Task Group 53.10), the rest forming their own Task Group, 53.3. There was shore gunfire to disturb the sweepers here, too, but again no mines were found, although a few were discovered near adjacent islands in the following days.

The last notable capture in this group of islands was Eniwetok. Here, the sweeper force consisted of two destroyer sweepers, two AMs and two YMSs. The same little group of ships was sweeping the landing craft in at each assault—only the division of ships between simultaneous attacks altered the pattern. Eniwetok gave the sweepers their first Japanese field to sweep, of 28 moored contact mines. This was quickly cleared that first day, the main hazards being fire from the shore and an equal amount from friendly ships! This lagoon became an important base and staging post for American warships moving forward to the later assaults.

Next, as a group of islands, came the Marianas. The capture of this group went on in parallel with the recapture of the Philippine Islands, with the available sweepers dividing themselves, as did the landing craft, between the two operations.

Saipan came first, in June 1944. For this assault the Minesweeping and Hydrographic Unit (Task Group 52.13) was split into six sweep units. The first two were composed of the old destroyer sweepers, the next two each had three of the newly-arrived AMs—*Chief, Champion, Herald, Heed, Motive* and *Oracle*, and the last two units each had five YMSs. The assault went in on June 15 and shore gunfire once again seemed to be the only hazard at first. The supporting warships were kept far offshore as a minefield was expected, and some of the sweepers had a rough time. Three days after the landings went in, YMS 323 received major damage as a result of hits.

The minesweeping force was increased soon after the main landings. There were a further eight YMSs in the Service and Salvage Group (Task Group 52.7), and in the last days of June a Japanese field was indeed discovered. However, it was on the other side of the island from the landing beaches, at Magicienne Bay, although further mines were found around the coast. The clearance sweep was not easy. Although the number of mines found was not large, they had been laid close to the coral reef, which meant tricky manoeuvring for the sweepers, at close quarters in the swell and currents with their wire sweeps out. In addition, there was plenty of fighting still going on ashore, quite close to them. The sweep was carried out by *Oracle, Chief* and three of the YMSs, and later many more mines were found ashore in sheds, ready to be laid.

Guam, five days later, gave no trouble, but the Minesweeping Group in the Southern Attack Force was sizeable—seven destroyer sweepers, four new AMs (*Skylark, Starling, Sheldrake* and *Swallow*) and six YMSs. Nevertheless, no mines were found.

Three days later again came Tinian, on July 24. In a week's assault, two AMs and two YMSs cleared all the waters surrounding the island, and cut 14 moored contact mines with little trouble.

The Palau Islands are part of that grouping, and three weeks later a landing was made at Peleliu, on September 15. A 45-strong force of minesweepers was included in the assault group, including all the old friends from the previous islands. Peleliu gave real trouble from Japanese mines, for the first time. The destroyer sweepers, going in first, discovered a big moored field off the landing beaches only two days before the assault, and nearly a hundred mines were cleared. In the process, the destroyer *Perry* hit one with her hull on September 13, and received severe damage, including a flooded engine room and nine killed. She sank soon after. On the day of the landings a fire support ship, the destroyer *Wadleigh*, hit a mine in the wake of a group of sweepers, and she too was severely damaged, although she was later salvaged.

Japanese mines were being swept around this island by the YMSs right through to the end of 1945, and YMS 19 blew up on one of them before the assault was finished, sinking quickly.

Another assault was taking place simultaneously with the landings on Peleliu, at Morotai, an important base to the east of Borneo. A small sweeping force there checked the waters, but found no mines, so the base was immediately put to good use.

Ulithi was the last important island in the Marianas to be captured, the landings going in on September 25 1944. There were extensive Japanese minefields, and the sweeping force had plenty of work to do. Destroyer sweepers, four AMs and a large group of YMSs carried out an intensive clearance sweep and, in a busy month, cut 230 moored contact mines. YMS 385 was lost to one on October 1, sinking quickly after striking it head on. In a side operation at Ngulu Lagoon, where a further field of about a hundred more contact mines was cleared, the faithful old destroyer sweeper *Montgomery* collided with a contact mine while at anchor, and was so seriously damaged that she was written off as a total loss.

Ulithi was to become one of the main forward fleet bases in the operations against Japan, and this assault was to pay handsome dividends.

The Philippine Islands now lay within the grasp of the American forces, and liberation operations were started without delay. The first assault went in in October 1944, but not until March 1945 were the islands to be once more in American hands. There were many gallant minesweeping actions during these protracted operations, and it is only possible to touch on them.

First came Leyte, where the assault commenced on October 20 1944. A large minesweeping force was included here; only one of the old destroyer sweepers, but no less than 13 of the new and very efficient AMs, plus 26 YMSs, showing that the stream of new ships from America was showing results. A further group of six destroyer sweepers was held separately on this occasion, as a high-speed clearance group.

The Minesweeping Group sailed from Manus for the Leyte assault on October 10. They needed to refuel at sea on October 15, from fleet tankers off the Palau islands. But the weather had become very bad, and the refuelling operation was really hazardous. The wind was blowing at over 30 knots, and there was torrential driving rain, so that no ship was able to complete her refuelling in under two hours. The YMSs were taking green seas solid over their decks and, during the night, they had to steam at full speed into those mountainous seas to keep up with the fleet. Many of them were struggling, and the fleet had to reduce speed to 9 knots to allow them to catch up. YMS 70 took so much water on board that she sank during the night, and the rescue operations by two of her consorts carried their own special form of hazard. Any navy man who has experienced the dangers and discomforts of refuelling in the open sea under gale force conditions will be able to picture the remarkable fight which this minesweeping force had to put up against the elements.

When they arrived in their sweeping area off Leyte on October 17 the weather was still atrocious—driving rain with very low visibility combined with mountainous seas—conditions hardly suited for handling wire s sweep gear in small ships. But the clearance sweeps were still carried out: first the destroyer sweepers ahead of the fast transports for the preliminary landings at the mouth of Leyte Gulf, then the AMs in two groups clearing a way into the gulf itself, cutting moored mines as they went. An APD (*Sands* this time) used her little LCVPs with rope Oropesa sweeps to check the water from the five-fathom curve into the beaches, and the assault went in on schedule.

The main moored field which had been found was the subject of intensive clearance sweeps during the following week, and by October 23 315 moored contact mines had been cut and destroyed. As Leyte was an important base, some of the AMs stayed behind for a while after the landing force had moved on, to check that all the mines had been found.

Ormoc Bay came next, with landings on December 7 1944. Eight AMs were included in this sweeping force, but no mines were found.

Mindoro followed on December 15, with a Minesweeping Unit of nine AMs, seven YMSs, and an APD with LCVPs, but no destroyer sweepers this time. The Japanese air attacks were beginning to make themselves felt now. *Saunter* and *Scout*, two of the AMs engaged some Kamikaze aircraft, and *Saunter* turned so fast under full helm that she hit a reef. It took a while to get her off, but no mines were found.

Then came Lingayen Gulf, on January 2 1945. Many mines were expected here, so another large minesweeping force was assembled—ten destroyer sweepers plus two DMs in support, ten AMs and no fewer than 44 YMSs. They sailed from Leyte and, on arrival, the destroyer sweepers cleared a passage into the gulf for the big fire support and landing ships. But they found no mines, and the main opposition came from determined Japanese air attacks. In one of these the destroyer sweeper *Long* received a direct hit and, suffering many casualties sank next morning. Some of the YMSs suffered near misses, but only minor damage—the attacks were not as effective as those at Okinawa were to be a little later.

The sweeping force retired to open sea for the night, as had become normal assault practice. While they were so doing, the Japanese made a torpedo attack on a group of the destroyer sweepers. *Hovey* was hit amidships and went down in just three minutes, with heavy casualties. Then another ship of the same group received very near misses from two bombs in another attack, and she also sank with heavy losses.

In all that sweeping under heavy air attacks, only three moored contact mines were cut, and those fairly close inshore to the beaches by the little YMSs.

Corregidor was next on the list, and bound to be a big one. The sweeper fleet was not all that large—just six AMs and 15 YMSs—but they finished up with a bag of well over 400 mines, a mixture of Japanese moored and controlled mines, and some of the old American moored controlled mines from the early war days.

On February 13 1945 the AMs opened play by starting an exploratory sweep, some three miles wide, inshore from the 100-fathom curve. They then moved inshore to within a mile of the beach, supported by fleet destroyers with 5-inch guns. This was where 28 of the old American-controlled mines were cut. Next day the AMs tackled a different area, and the Japanese defenders took violent exception, opening a hot fire from the shore. But the sweepers found mines too, cutting well over a hundred Japanese moored mines during that one day. Two supporting destroyers got too close in among the mines, and had a difficult job finding their way out again without hitting one.

The YMSs were tackling a different area, and again shore gunfire disrupted operations. Here, YMS 48 received a number of direct hits from a heavy calibre shore gun, was set on fire, and had to be sunk by destroyer gunfire as she could not be towed to safety in face of the shore opposition.

In yet another area, three YMSs tackled a field, supported by two more of the excellent fleet destroyers. Many mines were sunk, but this time both supporting destroyers struck and were damaged by mines, as was YMS 8: she caught a moored mine in her sweep which exploded while being cleared.

Meanwhile, the AMs had moved on and cut 270 contact mines in one area. Although there was little shore opposition, some YMSs were straddled by shore batteries, and one hit.

The widespread landings in the area had succeeded by mid-February, and the landing force moved on to the next task. However, more mines were known to be

Left *American YMS moving into echelon formation, as they prepare to commence an Oropesa sweep in the Pacific in 1945. The buoy boat is in the right distance.*

Right *The American Destroyer Minesweeper* Doyle, *of the 'Livermore' Class, with Oropesa sweep and LL reel aft, and the SA sweep amidships, with a derrick to handle it by the forward funnel.*

in Manila Bay, inside Corregidor, so a Manila Bay Minesweeping Unit was formed. This consisted of one AM, *Souffle*, leading 15 YMSs with some LCVPs for inshore work. The bay was a large area and, by Easter, a further 600 mines had been destroyed. The AM *Saunter* joined in for a time, but regretted it, as she hit a moored contact mine and suffered very heavy damage, but was sufficiently seaworthy to be towed away for repairs.

Mention should be made of two other landings before closing this epic chapter in the history of United States Navy minesweeping. At Zamboango, on March 8 1945, 12 YMSs swept in ahead of the landing force, coming under heavy shore fire close to the beaches, although no mines were found. Then, on March 18 at Negros, six YMSs had much the same experience. There were smaller operations going on within these hotly disputed islands throughout this period, with the YMSs especially going right inshore, and many fierce little actions were fought with the defenders, although mines were not always present.

The last island assault we should mention, indeed the essential preliminary to the assault on Okinawa, was the capture of Iwo Jima. The landings commenced on March 18 1945, and 35 minesweepers were present—16 AMs, six of the new 'Livermore' Class fast destroyer minesweepers and 13 YMSs with the Senior Officer aboard the new cruiser-minelayer *Terror*, acting for the first time as minesweeper flagship.

The waters around this island were deep, and there were no natural harbours, but check sweeps were carefully carried out. There was a good deal of enemy shore fire, but no mines were found. However, the old destroyer sweeper *Gamble* received a direct hit from a Kamikaze, and had to be written off.

Assault on Okinawa

This great assault must rank alongside the invasion of Normandy the previous summer as one of the biggest and most difficult landing operations of World War 2.

From the naval viewpoint, far fewer warships—including minesweepers—were engaged than in the landings on the northern French coast, but losses

of ships from what might be termed normal enemy activity were also lighter. However, Okinawa produced enormous logistic problems, and the great Kamikaze campaign was a challenge requiring even greater feats of courage and of ship repair.

After the capture of Iwo Jima, Okinawa was the last Pacific island held by the Japanese. It lay directly on the American route to Japan—from Okinawa, fighter-escorted bombers could reach the Japanese mainland, and from the island a great, final seaborne assault could be mounted against Japan itself. But it was not just for this reason that Okinawa was held in awe; it was also for the sheer logistic difficulties in taking it.

From Hawaii (the main American fleet base in the Pacific), Okinawa lay 4,040

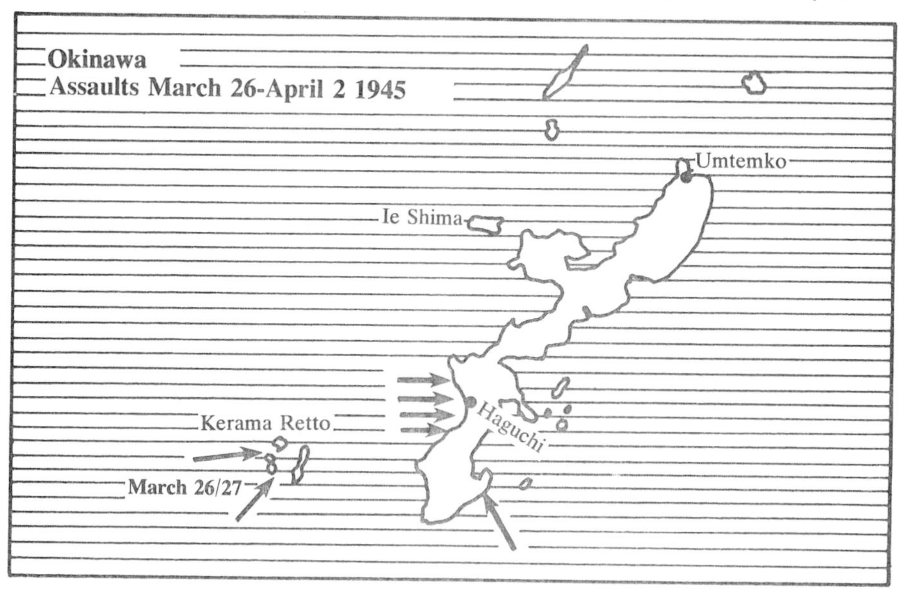

miles distant; even from the captured forward base of Ulithi it was 1,200 miles, and 1,400 from Leyte. Compare this with the 75 miles or so from Spithead to the Normandy shore. Extensive preparations were thus necessary before the assault itself on April 1 1945.

Even before all of Iwo Jima had fallen to the Americans, warships and landing craft were being withdrawn to prepare for Okinawa.

The Joint Expeditionary Force of 1,213 ships included a Minesweeping Flotilla of 116 vessels, which are listed at the end of this section. They assembled at Ulithi in the first half of March. This flotilla showed the results of the enormous shipbuilding programme which had been driven through in the United States during the previous three years; all but three of those 116 ships were newly-built—28 of the new 'Livermore' Class fleet destroyers, with a full outfit of minesweeping gear (the Royal Navy kept all of its fleet destroyers in their primary role, due to shortage of ships); 43 of the two new classes of fleet minesweeper (AM), and 41 of the excellent little yard motor minesweepers. The flotilla was under the command of Rear Admiral Sharp, flying his flag in the new cruiser minelayer *Terror*, and the whole was designated Task Group 52.2 within Task Force 52.

Preliminary operations

The minesweepers left Ulithi in two groups during March 19, and headed for Okinawa—Operation 'Iceberg'. It was not an easy passage—one destroyer minelayer and three fleet minesweepers had engine trouble, but they all made it. At dawn on March 25, Love Day minus eight, they arrived off the islands, ready to start minesweeping operations.

The island of Okinawa runs north-east–south-west, and some 20 miles off its western foot lie the smaller islands of Kerama Retto. The assault plan called for the capture of these first, and then for the main assault to go in on the western side of Okinawa, at a large bay called Haguchi.

So, on March 25, the sweepers began their mission in both areas. The first task was to clear an area to ensure a safe approach for Task Force 52, the Amphibious Support Force, for Task Group 51.1, the Western Islands Attack Group, and for Task Force 54, the Gunfire and Covering Force.

During the first day the sweepers were specially covered against Japanese shore gunfire by the battleships of Task Force 58 (the Fast Carrier Task Force), including the famous USS *New Jersey*, while fighter cover for them was directed from *Terror*. The supporting gunfire was of necessity at long range, but after the sweepers had completed their first search, the supporting ships could move closer inshore for visual gun control, and the general fighter air cover was subsequently sufficiently effective to cover the sweepers.

The objective of the sweepers was to clear no less than 2,500 square miles of sea before the main assault went in. Their gear was set to sweep for moored mines down to a depth of 60 feet, as well as for magnetic and acoustic mines. For these influence mines, they would sweep from the 100-fathom contour, in as far as their respective draughts would permit them to go. The weather, fortunately, was fine—the wind was easterly at 13 knots, and there were no waves more than two feet high.

The sweepers were led in by two units of destroyer minesweepers, carrying out a fast sweep, first to the south and south-east of Okinawa, and then westwards towards Kerama Retto. Then the units of fleet minesweepers (AM) came in, searching the waters south of Kerama Retto to clear the way for the assault on those islands. They then continued on to clear the approaches to Haguchi, for the deep-draught supporting warships, attack transports and LSTs.

For six days this intensive minesweeping operation was maintained with no real interference from the enemy, and 123 mines were swept—all the moored contact type. During the final two days before the assault, the battleships closed the range and engaged in visual counter-battery fire.

At one stage, on March 27, the sweeping operations dropped a little behind schedule, but the immense area was still swept in good time for the assault. As with the Normandy landings, there were sweeper casualties before the landing force appeared on the scene. Two destroyer minesweepers, *Adams* and *Dorsey*, received major damage in Japanese air attacks. *Adams* fought off several Kamikaze pilots before one hit her stern, and *Dorsey* also received a direct hit; both had to return for repairs over all those long miles to Navy Yards in the United States. The fleet sweeper *Success* was able to report that three torpedoes from a Japanese submarine had missed her. But the fleet sweeper *Skylark* hit a mine off Haguchi, while a field discovered there was being cleared, and burned fiercely before sinking. The fleet destroyer *Halligan* was also mined, and her wreck drifted ashore; the fleet sweeper *Breese* carried out a courageous rescue mission at night to pick up her survivors.

Then the sweepers moved further inshore again, to clear channels off the beaches for the fire support ships, and anchorages from which the attack transports could launch their landing craft. Swept channels were also needed right inshore for the larger LCIs and the even larger LSTs which would follow them, and now the smaller YMSs were in close action, with destroyer minesweepers providing close fire support with their 5-inch guns. During all these intricate operations, the minesweepers were the closest ships to the enemy, before the landing forces went in.

By the last day of March all 2,500 square miles of sea had been cleared by the sweepers, and 222 mines had been destroyed. As was the case right through the assault period around this hotly disputed island, many floating mines were sighted and destroyed by gunfire, and it was thought that the Japanese were launching these from the beaches by night.

The assaults go in

On March 26 landings were made on schedule on Kerama Retto; they were a necessary preliminary, as these islands would be badly needed as a repair base for warships damaged in the forward areas. Indeed, four dry docks were specially towed over from the United States and stationed there, and the repair organisation carried out untold feats as ships damaged by suicide planes came in one after the other. Even some of those with major structural damage were repaired in this forward area, rather than sending them all the way back to America. This had a marked effect on the strength of the naval forces at any one time around the island, however fierce were the Kamikaze attacks.

Then, on April 1, the main assault went in at Haguchi. The sweepers had started their daily clearance operations, going right inshore to clear the boat lanes (where the very smallest sweepers were invaluable), and the underwater demolition teams had cleared the beach obstacles. So the first wave of 20,000 men went in, and the sweepers heaved a great sigh of relief.

It is interesting that, from this point on, as many floating mines were sunk by gunfire from various ships as were cut by the sweepers themselves. But the daily sweeps were carried out without a break, and further clearance, around the coasts of Okinawa, was pushed ahead with great determination.

Mines were still being cut around the southern tip of the island as the main assault went in, and 26 were sunk by gunfire from boats close inshore. Some were even sunk by strafing gunfire from the carrier aircraft as they flew low over the assault area, and hardly a day went by without the bag of swept mines going up by one or both methods.

Now the swept channels were being marked more clearly (again following the Normandy pattern); one-way ship channels were marked by light buoys with radar reflectors, between Haguchi and Kerama Retto, and extended as the assault progressed ashore.

The Kamikaze attacks continue apace

The Kamikazes were attacking in great fury. The AM *Skirmish* sustained minor damage on April 2 when she took a near miss, and three days later the destroyer sweeper *Hambleton* also received minor damage.

Then, on April 6, the suicide planes launched their biggest attack of all, and in one week 355 Kamikazes, accompanied by 350 other aircraft, attacked. Six American ships were sunk, ten more written off as constructive total losses and a further seven heavily damaged.

Two of the destroyer-type sweepers, *Emmons* and *Rodman*, fought a famous battle. During the afternoon they were attacked by a large number of suicide planes, and put up a furious defence with their excellent armament of 5-inch guns, multiple Bofors and Oerlikons (although the latter proved less effective than the larger guns against these attackers). First *Rodman* took two near misses, then a Kamikaze crashed into her bridge and the ship was wrecked forward. Her crew displayed great damage control skills, and during the night she made it back to the repair base at Kerama Retto, but had to be sent back to a Navy Yard in America for major structural repairs.

Then *Emmons* fought off numerous attackers, bringing down six or seven Kamikazes herself; but her luck was too good to last, and finally, off the north-west coast of the island, no less than five suicide aircraft managed to dive directly into her, destroying 30 feet of her stern. The burning wreck of this fine ship had to be sunk by her consort *Ellyson,* as it drifted inshore towards the enemy-held beach.

Other sweepers were also in trouble that day. The destroyer minelayer *H. F. Bauer* received minor damage in an air attack, and the fleet sweeper *Devastator* was hit in the forward engine room by a suicide plane. Despite heavy hull damage, she had only one hole below the waterline and remained operational; Kerama Retto was able to patch her up later.

The destroyer *Leutze* received a hit, and the fleet sweeper *Defense* took her in tow, only to be grazed herself by two Kamikazes at nine in the evening—but she carried on with her mission. Other ships receiving air attack damage that day were the fleet sweepers *Facility* and *Ransom,* and YMSs 311 and 321; a Kamikaze crashed into the sea off the bow of YMS 311, glanced off her stern, damaged the 3-inch gun and caused fragmentation damage forward.

But sweeping takes priority

The daily clearance sweeps continued without a pause, and the bag of mines swept grew. It was thought that the Japanese were laying new minefields by night, so exploratory sweeps were carried out each morning in suspected areas; and on the day after the assault 24 mines were swept in this way in one field, off the east coast of Okinawa.

Again following the Normandy pattern, small Japanese battle units made their appearance at night from the shore; Sweep Unit 2 happily tracked a group of them one night, all moving at better than 30 knots, but they did not come out and attack.

On the evening of April 7, YMS 427 (one of the latest units of the class) was hit by a Japanese coastal gun, which was promptly silenced by the heavy cruiser *Wichita,* cruising nearby in support. Then, in the big bay on the eastern side of the island, Nakagusuku Wan (which was to give the sweepers a great deal of trouble), PGM 18, one of the inshore supporting force of gunboats, hit a moored mine and sank quickly, losing many of her crew, while YMS 103, one of the sweep unit which PGM 18 had been supporting, turned back to pick up her survivors. But it did not end there, for YMS 103 herself then struck two mines, both in the forward part of the ship, and the whole hull was blown away to a point aft of the bridge. She was beached, in case she could be salvaged later, but lost four dead and 12 wounded from her small crew. Later that afternoon YMS 92 struck a mine under her stern in the same field. She lost her stern, but luckier than YMS 103, was towed back to the repair yard at Kerama Retto. Nineteen mines were cleared from the field that day.

The second week of the assault saw some easing off in the Kamikaze attacks, but they still continued.

On April 12 a crashing Kamikaze dealt the fleet sweeper *Gladiator* a glancing blow, luckily with only slight structural damage, but the same day the destroyer sweeper *Jeffers* suffered major damage forward. Then on April 16 two more destroyer sweepers, *Harding* and *Hobson,* also suffered severe Kamikaze damage and had to return to American Navy Yards for major repairs. These three actions reflected the very exposed position of the destroyer radar pickets, stationed far out from Okinawa to warn the fleet of approaching air attacks. The destroyer sweepers, newly built with good radar and heavy anti-aircraft armament, were used to reinforce these pickets, and so suffered heavy losses too.

The fleet sweepers were used as units of the anti-submarine patrols maintained round the islands, to protect the landing forces. The larger 'Auk' Class joined the outer screen, and the smaller, slower 'Admirable' Class vessels were deployed in the inner screen. This protection was considered so important that many of the fleet sweepers were withdrawn from sweeping for this monotonous work, leaving the remaining fleet sweepers with the YMSs to carry out the daily clearance sweeps inshore and in the channels.

A side landing was made by the US Marines on the smaller island of Ie Shima off the west coast of Okinawa, and Sweep Unit 10 carried out an exploratory sweep before the assault, although no mines were found.

But the crowded area of shipping inshore inevitably suffered too. At dawn on April 10 the destroyer sweeper *Hambleton* and YMS 96 collided, four miles offshore, and the Yard Minesweeper was towed away to Kerama Retto with its engine rooms flooded.

Losses from enemy action continued during the last week or two of April. On the 19th the fleet sweeper *Spear* took minor damage forward, from yet another Kamikaze which crashed and exploded close to her port bow, causing fragmentation holes. Then, on the 22nd, the fleet sweeper *Swallow* took a direct hit in her engine room from a Kamikaze, close to the crowded Kerama Retto repair base, and sank in just a few minutes with heavy casualties.

April 27 saw another Kamikaze hit the destroyer sweeper *Butler*. She was one of the ships on detached radar picket duty, and although with excellent gunnery she shot down several attackers before they could get to her, she took a near miss from one which caused major damage aft, including her sweep gear. However, the efficient forward repair base was able to keep her in service. The last sweeper to be damaged in April was the destroyer *H. F. Bauer*, suffering her second Kamikaze hit of the assault period.

By the end of the first month, the whole of the north-eastern side of the island had been cleared by the sweepers, as far out as the 300-fathom contour, which was a magnificent achievement in the face of such heavy opposition. But the strain on the crews had been very great.

The second month of assault

The battle for Okinawa was settling down now. Ashore it was a war of attrition against the entrenched Japanese defenders. But at sea, the Kamikaze campaign continued without a break—not quite so hectic as in the first few weeks of the assault, but every hit was just as damaging.

The first major casualty occurred on the very first day of May, and was to the Minesweeper Flagship, the new cruiser minelayer *Terror*. She was flying the flag of Rear Admiral Alexander Sharp, now carrying the title of Commander, Minecraft, Pacific Fleet. Once the initial assault had been completed successfully, and the islands of Kerema Retto secured, she was stationed in the repair base, ready to look after the damaged minesweepers as they came in from the battle. Before dawn on May 1 a solitary Kamikaze came in over Kerama Retto, flew past *Terror*, then turned, came back and struck her midships superstructure. The aircraft carried two 500 lb bombs: one exploded as it hit the ship, the other penetrated 18 feet before exploding. Then the aircraft's engine ripped through the decks, finishing up in the wardroom. The hole where the plane entered was relatively small and neat, but the havoc caused inboard was considerable. *Terror* lost 41 dead, six missing and 123 wounded in that action, and had to withdraw from the battle, retiring to San Francisco where she arrived a month later for repairs. Among the dead was Captain Robley W. Clark, Commander of Mine Squadron 1, and a valuable senior officer on Admiral Sharp's staff.

The Admiral and his staff, after some hair-raising adventures, transferred to the US Coast Guard cutter *Bibb* which was lying near by, and on board which, as it happened, Admiral Sharp and some of his staff had been planning the minesweeping side of the final assault on Japan.

Two days later there were two more major casualties, both again destroyer minesweepers—*Macomb* and *Aaron Ward. Macomb* took a Kamikaze hit on deck and suffered major structural damage, although the splendid forward repair base was able to patch her up. *Aaron Ward* was another matter. In the evening of May 3, about a hundred miles west of the main island, she was jumped by a large group of suicide planes. Her heavy armamant helped her to fight them off, but ten made a dive specially for her and five scored direct hits, while three others either scraped the masthead antennae or crashed into the sea close enough to hurl debris aboard. The ship became a mass of flaming wreckage, but with great heroism her crew, with the help of the destroyer *Shannon*, got her back to the repair base at Kerama Retto. *Aaron Ward* lost 19 dead, 20 missing and 39 wounded in this attack, and was awarded the Presidential Unit Citation.

Next day, several more sweepers were in trouble. The fleet minesweeper *Gayety* was damaged (for the third time) by a Kamikaze but her damage fortunately was minor, and the repair base soon put her back into the battle. But one of the three 'four-stacker' destroyer sweepers, *Hopkins*, received a very near miss from a Kamikaze, and was lucky to be still afloat. She was supporting a group of YMSs at the time, west of Okinawa, and this was one of a number of attacks which disrupted sweeping on May 4. Three of the YMSs were damaged, too—331, 327 and 311. Two of these also received hits from friendly gunfire, during the hectic attempts to shoot down the Kamikazes before they could find their targets.

The daily search sweeps continued monotonously, and on most days a few mines were destroyed, the majority of them still being found floating.

On May 10 a special base anchorage was commenced for the minesweepers, in Untemko, *Hopkins* and YMS 319 surveying the bay and laying channel marker buoys. This saved many weary journeys between sweeps back to Kerama Retto.

May 25 saw a Kamikaze making a near miss on the fleet sweeper *Spectacle*, and two days later, in the early morning, no less than 25 aircraft dived on the destroyer sweeper *Forrest*, one of which made a direct hit, causing major structural damage.

Then, on June 21, 82 days after the original landings, Okinawa was finally declared to be entirely in American hands, and the sweepers could count the cost.

Out of 25 American warships sunk during the battle, four were sweepers—one DMS, two AMs, and one YMS. Of 131 warships damaged, 16 were sweepers—nine DMSs, five AMs, and two YMSs. This list excludes those with only minor damage. But no fewer than 510 mines had been swept, all of them, as far as records show, of the moored contact variety, with many found floating. It was further estimated, as far as one could be sure during the fierce anti-aircraft battles, that the sweepers had shot down 95 aircraft, with half as many again as 'probably destroyed'.

Onward from Okinawa

Plans were well advanced by this time for the final assault on Japan, and the task of the minesweepers, as always , would be to be up there in front clearing the way.

During the carrier aircraft raids on the Japanese mainland, the sweepers were not involved, as the fleet was operating far offshore. Most of the capital units streamed paravanes as a precaution, but no mines were cut and no damage suffered.

But before Japan surrendered, after the two atomic bombs had been dropped, the US Fleet's minesweepers had three more operations to carry out.

The first—code-named 'Zebra'—was to carry out a massive search of the sea stretching from Okinawa into the East China Sea. This was not a major operation—some two dozen sweepers, with supporting ships to help ward off air attacks. But about 2,000 square miles of sea were swept in the middle weeks of June, and over a hundred moored contact mines were cut.

The second—code-named 'Juneau'—took them a major step towards the Japanese mainland, This time, the area to be cleared was no less than 9,000 square miles, and 100 sweepers (as many as in the main assault on Okinawa) were employed. Reports describe how the sweepers were deployed in a single formation in echelon, sweeping a path six miles wide as they went. Battleships were in support, and the carriers flew intensive patrols over the long formation. Four hundred moored contact mines were cut in that operation.

Lastly came a further sweep towards the Japanese mainland, again strongly supported from both sea and air. This operation was code-named 'Skagway', and again nearly a hundred sweepers were engaged in it. They sortied from Okinawa in early August, and although Japan surrendered the day after they reached their sweeping area, the minesweepers continued without a break. Their eventual bag was over 650 moored contact mines in that remarkable operation.

The final scene took place in Tokyo Bay, as the US Third Fleet accepted the Japanese surrender. Five fleet sweepers of the 'Auk' Class led the fleet in, and a representative group of DMSs, AMs and YMSs were present at that great moment. But the ordeal of the sweepers—even around Okinawa—was by no means over.

Minesweepers for Okinawa

It has not been easy to find exact records of the sweepers present at the assault on Okinawa, but what follows should be reasonably accurate; certainly the numbers of each type seem clear, but the composition of the individual sweep units may have varied somewhat, especially after the initial sweep.

Destroyer Minesweepers (DMSs): 14 'Livermore' Class ships, completed in 1944, and disposed in three sweep units (Nos 2, 3, and 4).

Fleet Minesweepers (AMs): 25 'Auk' Class plus 18 'Admirable' Class, disposed in ten sweep units.

Yard Minesweepers (YMSs): 41 ships of the class, probably disposed in six sweep units.

Destroyer Minelayers (DMs): 14 ships of the 'Livermore' Class, one accompanying each of the sweep units; their role was probably a multiple one, including sinking floating mines cut by the sweepers, close fire support while free to manoeuvre, and transport of spare sweep gear.

Destroyer Minesweepers (DMSs): three ships of the old World War 1 'Dorsey' Class (the 'four-stackers'). Their role in this operation was probably much as that of the Destroyer Minelayers.

Minesweeper Flagship (CM): *Terror*, one of a new class of three.

Support Gunboats (PGMs): 16 of this class, for close inshore gun support.

Summary of minesweepers

14 'Livermore' Class DMSs
25 'Auk' Class AMs
18 'Admirable' Class AMs
41 YMSs.
98 ships in all

In support
14 'Livermore' Class DMs
3 'Dorsey' Class DMSs
1 M/S Flagship
18 ships in support

Total minesweeping force—116 ships

Minesweeper casualties at Okinawa

Sunk: One DMS, two AMs and two YMSs—total five ships. Damaged: One M/S Flagship, eight DMSs, two four-stack DMSs, seven AMs and six YMSs —total 26 ships.

YMS 478 ashore at Okinawa after one of the great typhoons which hit the sweepers in 1945. This picture should be seen in conjunction with the listed YMS losses.

The British Pacific Fleet for Okinawa

It was at the battle for Okinawa that the British Pacific Fleet, with its fleet carriers and battleships, joined the US Fleet for the first time in full force.

The BPF was allocated targets in Sakishima Gunto, the islands between Okinawa and Taiwan, and the fleet launched air strikes from the sea at the airfields there throughout the Okinawa campaign.

From the minesweeping point of view, it should be remembered that the Fast Carrier Task Force, operating far offshore, did not need sweepers with it—the big ships streamed paravanes, as did the American carriers and battleships. There were no mining incidents, but nevertheless, the BPF had minesweeping units allocated to it and, although these were not operating at full strength by VJ-Day, they were building up fast.

The first sweepers on the scene, clearing the waters in and around the fleet bases and protecting the Fleet Train as it entered hostile waters to replenish the Fast Carrier Task Force, were 18 Australian fleet sweepers of the war-built 'Bathurst' Class. They would have been supplemented by two full flotillas of 'Algerine' Class fleet sweepers from the Royal Navy, with four special 'Isles' Class danlayers attached to them, and a further 'Algerine' Class unit converted to an M/S Headquarters Ship, would have been with them, together with a specially converted M/S Store Ship. Part of the BYMS force was also allocated to the Far East, but it is probable that most of these would have been retained in the East Indies Fleet, which had far more widespread assault operations to launch than did its Pacific counterpart.

Chapter 8

Postwar mine clearance

Great as were the minesweeping operations in all theatres of war during the period of hostilities, no less important were the mine clearance operations over the following few years, in which the minesweeping forces of all nations worked together.

Northern Europe

As soon as hostilities ceased on May 8 1945, energetic steps were taken to tackle the enormous problems in clearing all the waters, usually crowded with merchant shipping, around the North Sea and the English Channel, together with other special areas such as the Baltic.

The Admiralty coordinated this operation and set up an International Mine Clearance Board to handle the organisation. All available sweepers of all nations involved were pooled for the task; German and Italian sweepecs came under the control of either the Royal Navy, or that of the country round whose shores they were sweeping.

The Royal Navy contributed some 300 minesweepers for this operation alone—100 fleet sweepers and 200 of the wooden MMSs, nearly all of which had been built during the war. These sweepers were distributed around the British coast as follows, and this chart also shows the number of mines suspected in each area:

UK port	Fleet sweepers (flotillas of 8)	MMSs (flotillas of 10)	Mines Moored	Mines Ground
Dover	1	2	83	2,469
Sheerness	0	3	0	2,750
Harwich	3	0	465	0
Lowestoft	0	3	0	2,812
Grimsby	1	4	275	3,840
Edinburgh	2	1	0	2,512
Newcastle	1	1	183	0
Scapa Flow	1	0	255	0
Aultbea	1	0	403	0
Milford Haven	1	2	213	3,840
Liverpool	0	1	0	3 443
Plymouth	0	1	78	346
Portland	0	1	67	916
Portsmouth	1	2	74	1,838

YMS 438 returning to San Francisco in 1945, after distant service in the western Pacific. Her radar aerial, as in many cases, no longer carries its cover.

These were only for clearance of the mines around the British coasts, yet it was estimated that it would take 549 days to clear the moored mines and 676 days for the ground mines.

In addition, British sweeping flotillas were working around the coasts of France, Belgium, Holland, Norway and Germany. With them were working the MMs transferred to the Allied navies over the war years. Cooperation was magnificent, but the sheer volume of mines to be swept meant that only the previously swept channels could be maintained for the time being.

Clearance sweeping would widen these channels gradually, so reducing the risk to the merchant ships using them. Where unsweepable ground mines were known to be present, however, it was thought better to let the mines die a 'natural' death—as silt covered them, or their perishable internal fittings came to the end of their life—rather than risk sweepers unnecessarily in peacetime.

The North Sea was one of the most heavily mined areas. First, the original British mine barrier, stretching offshore right down the East Coast, from John O'Groats to the Thames, had to be cleared meticulously. Only moored contact mines had been laid here, and the winter gales of each war year had taken their toll of the mines which had come adrift, many being washed ashore. But flotillas of 'Algerine' Class fleet sweepers, each with their attendant danlayers, carefully swept every mile of the area. Relatively few mines were cut and destroyed, but none was left. Accurate navigation was essential and taut wire measuring gear, from the nearest headlands, was much in use.

Then the dangerous North Foreland channel, from Ramsgate to the River Scheldt, was cleared as far as possible. Here, pressure mines were suspected, and it was in this channel that the remarkable 'Egg Crates' had their only operational use. Designated the Fiftieth Minesweeping Flotilla, the 'Egg Crates', with their attendant 'Bangor' Class tugs and other small craft, were based at Ostend, moored to specially laid buoys outside the harbour. They carried out check sweeps of the channel regularly, and in June 1945 EC 10 swept one ground mine, lifting it 30 feet astern of her in nine fathoms of water, north of NF 7, the buoy which was surrounded by the wrecks of mined ships. At the time, the swell recording vessel was registering zero, and it was thought that the mines were lying between ridges of sand. They were certainly hard to sweep. Another 'Egg Crate', EC 7, came adrift from her tugs in a full gale that autumn, and was wrecked on the sandy shore near Blankenberghe.

Some of the early mine location units used in Europe were tried out in diving to locate ground mines in this channel, although the water was so muddy, with fast tides, that their work was difficult.

A little further west, the ships of the Steam Gunboat Flotilla were in use as fast LL/pressure mine sweepers. *Grey Shark* and *Grey Seal* had their first success in July, lifting a red magnetic seven miles north of Boulogne, and got three more near by later in the day.

Two flotillas of fleet sweepers were also working in the North Foreland channel and in May 1945 alone they lifted 51 ground mines and cut nine moored.

'Algerine' Class fleet sweepers deployed in quarter line, as they prepare to stream Oropesa sweeps during postwar mine clearance operations. This was the 3rd Minesweeping Flotilla, led by Bramble, *with* Mandate, *the Second Senior Officer's ship, in the foreground.*

During the same period the BYMS and MMS flotillas still based at Ostend and Terneuzen, lifted 170 ground mines and cut 21 moored.

The approaches to the Dutch ports also needed to be cleared. Force A, of 18 MMSs, six M/S MLs and two HDMLs, sailed from Sheerness the day after VE-Day to tackle this area. The 'Algerine' Class fleet sweeper *Prompt* also sailed with them, as their HQ ship, but was mined in the NF channel on the way over, and had to return to the UK. Towards the end of the year, the Third Flotilla of 'Algerines' was operating off the Dutch coast, and had great difficulty in sweeping mines in a sea which was full of ice floes.

In the Nore Command, the number of mines swept in the first half of 1945 was the highest since the grim days of early 1942, and the second half of 1945 was not much different.

The trawler *Rolls-Royce*, still in faithful naval service, lifted her 197th ground mine on July 23; she was one of the last fishing trawlers to leave the Navy, and had performed noble service.

Casualties continued, too. Merchant ships, and some sweepers, were to be sunk or damaged for several years yet. In 1949, for instance, the Belgian cross-channel steamer, *Prinses Astrid* was sunk by a ground mine which exploded right underneath her engine room, just outside Dunkirk harbour. And this was in a channel carefully swept every day, and used by many ships. The mine was thought to be a magnetic, buried under sand, but uncovered suddenly by a shift in the tides and currents.

Off Germany and Norway, similar intensive weeping efforts were continuing. All the German 'M' and 'R' classes of sweepers were in service under British control, numbering 19 flotillas and 379 ships in all. A swept channel was quickly completed from the Humber to the Heligoland Bight, and German charts were closely examined to find the positions, not only of German defensive fields, but also the fields of aircraft-laid Allied ground mines. Many of these could not be swept, and were not equipped to sterilise themselves. In the Heligoland Bight itself, British 'overlap' mines had been laid a year previously and had not responded to German sweeps; now they were equally unresponsive to British sweeps, although some forms of SA sweep could render them inactive.

Round the coasts of Norway, too, there were many mines to be swept. Four 'Bangor' Class fleet sweepers were loaned to the Norwegian Navy, and in the two months preceding July 1945, minesweeping forces in that area cleared 735 moored mines alone, with no less than 326 sweep obstructors. The mine barriers in the coastal channels had to be cleared before the winter gales began, and German sweepers were pressed into service there, too, *Sperrbrecher* No 23 being sunk.

The Ninth Flotilla of 'Algerines' went across from Harwich, and on return told a tale of sweeping for 607 miles in 78 hours, without once recovering their sweeps. This was thought to be one of the longest continuous sweeps on record, and had been carried out in the constant light of the midnight sun. Two other flotillas of fleet sweepers were also operating off Norway and, in addition, there were 50 MMSs or similar vessels, eight M/S MLs and many trawlers.

Down the English Channel, similar intensive sweeping operations were under way. The number of MMSs manned by French seamen grew, and they also operated some YMSs. The Canadian Thirty-first Flotilla of 'Bangor' Class fleet sweepers was retained in European waters for quite a long period, doing excellent clearance work due to the shortage even then of fleet sweepers. Demobilisation was by now making the manning of this big sweeper fleet increasingly difficult, and many of the crews were coming from the frigates and corvettes which were going into well-earned reserve.

In the Channel Islands, the only part of the British Isles to be occupied by the Germans, a fleet sweeping flotilla composed of MMSs and M/S MLs swept in the liberating force.

Right through the second half of 1945 and into 1946, an average of 50 ground and 350 moored mines was swept each week in the northern area of Europe.

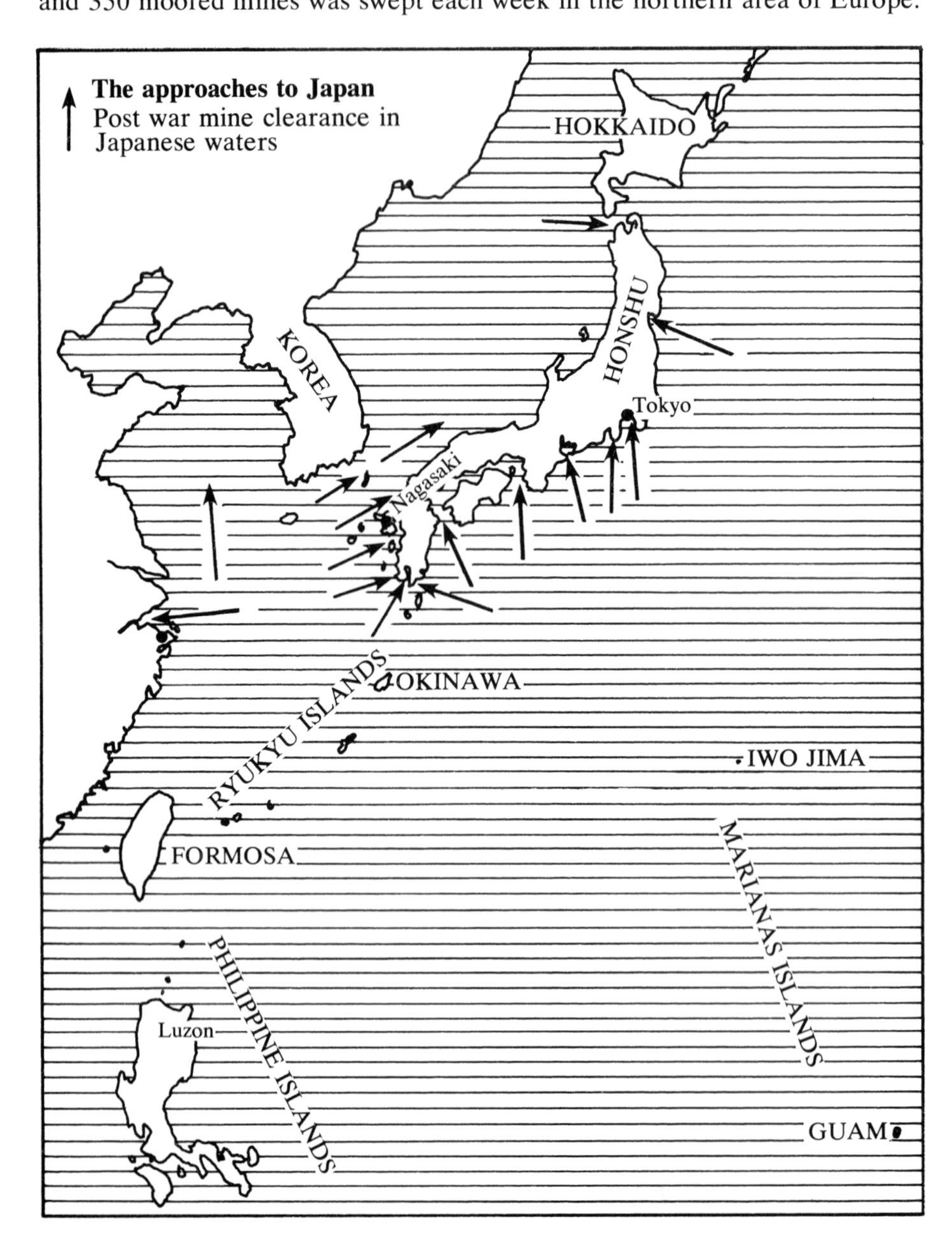

Swept channels were kept carefully buoyed, and special charts and route instructions were issued regularly.

In early 1952, the International Mine Clearance Board was disbanded, and each national authority became responsible for ensuring that ships were not mined in its waters.

In the Royal Navy, the BAM Class fleet sweepers and the BYMSs had long since been returned to the United States Navy, or turned over on its behalf to other Allied navies. By 1949, only two flotillas of fleet sweepers (already by then redesigned 'ocean' sweepers) and a handful of MMSs remained in service. The last sweepable field around the British coasts, in the Thames Estuary, had been cleared in October 1948. Many of the fields of influence mines were left alone to die a natural death, and the swept channels, although widened to four miles in width, remained for some time.

A new generation of coastal and inshore sweepers, developed respectively from the MMSs and M/S MLs, was coming along, but the Royal Navy built no more ocean sweepers after the excellent 'Algerine' Class.

The Mediterranean

Here, British minesweeping efforts continued at a great pace, with Italian and Greek sweepers playing their full part.

The Second Flotilla of 'Algerines', with its Senior Officer in the HQ ship *Fierce,* was operating in these waters right through to 1950, and the number of mines lifted or cut was comparable to that in Northern Europe.

Mines had been laid thickly in the Adriatic, around the Greek islands and west of Italy, and the clearance operations were widespread and hazardous—the photograph of the fleet sweeper *Regulus* sinking at speed off Corfu illustrates the point.

Other British sweepers operating in the Mediterranean in 1945–46 included four flotillas of fleet sweepers (and the US Mine Squadron of six 'Raven' Class ships), plus several flotillas of BYMSs (some manned by the Greek Navy) and one of M/S MLs which were doing useful work in the widespread areas of inshore shallow waters. Walrus flying boats, useful with their slow, lumbering speed, and US Navy blimps were being used for mine-spotting in the clear water ahead of the sweepers round the coasts of Italy. HDMLs were active similarly in the Aegean, while the wooden 'ZZ' lighters were operating in the Venice canals, where they detonated three magnetic mines in June 1945.

The Pacific Ocean

When peace came with VJ-Day, a huge minesweeping operation was already under way in the approaches to Japan, and in completing the clearance of the captured islands—not least Okinawa—to give a safe approach for the planned final assault on Japan itself.

Now a mine clearance operation, using the full resources of both American and Japanese sweepers, was launched quickly in order to get all Allied prisoners of war out and home, and to get the occupying American forces into Japan.

As in Europe, the plan allowed for the clearance of all water essential for the movement of ships, but the avoidance of all other areas—especially where there were known to be fields of ground mines which could not be swept easily, or not swept at all.

The Australian and British sweepers present in the south-west Pacific concentrated on that area; the Indian sweepers were largely used in escort work; American YMSs, now in the operational area in great numbers, ranged far and wide through the islands, supplementing other sweepers or carrying the whole burden themselves.

The clearance around the Japanese mainland was divided into 11 different areas, round ports or headlands, and both moored contact and magnetic and acoustic ground mines were cut and lifted in large numbers. But as in Europe the totals of mines known to be laid did not tie up with the totals swept—many moored mines had broken away to sink, or to become dangerous floaters, while many ground mines became silted over, sterilised themselves, or just went dead.

Here, space only allows us to list some of these impressive clearance operations, all faithfully recorded at great length in the US Navy's combat records.

Tsugaru Straits, at the northern tip of Japan, yielded 920 mines to 25 sweepers.

Sendai-Shoshi, coming down the east side of the mainland, gave up 264 to ten sweepers.

Tokyo Bay saw the clearance of 77 mines to 23 sweepers.

Omai Saki only revealed 13, but the Japanese sweepers had already cleared 229 others.

Nagoya needed a big force of sweepers—no less than 72, mostly American—as there were many influence mines there. But only six ground mines were lifted.

Wakanoura gave up 385 mines to 52 sweepers.

Kobe held a big field of US influence mines which had already claimed many victims; but in a protracted clearance operation, 75 sweepers could only find two of them. It was hoped that the others had sterilised themselves.

Kochi-Shikoku was cleared of 220 mines, to a comparatively small force of ten sweepers.

Hiroshima contained another big field of American influence mines. But they sterilised themselves well, and only 30 were lifted during a very long clearance operation by a strong force of 45 sweepers.

Bungo Suido was another big one, a mixed field of moored and ground mines, containing about 3,500 of varying types. A clearance fleet of 90 sweepers tackled it, and cleared nearly 2,000 of them.

Van Diemen Straits, at the southern tip of Kyoshu, had another large field of 800, but again the clearance bag was very small, 11 sweepers lifting only 80 of that total.

Kagoshima was next on that southern tip, with a smaller field, but of the mixed American and Japanese mines 250 were cleared here.

Kadoura next door was an easy one—a small field, and only 75 mines were swept.

Nagasaki had received the attention of the American minelayers—700 mines—and 280 were cleared by ten sweepers.

Sasebo, close by, had a large field of well over 1,000 mines, and 65 sweepers cleared 460 of them.

Fukuoka, just around the corner on the west side of the island, had a large influence field. Fifty-four American and Japanese sweepers took four months, carrying out a careful check sweep, to make sure that all were sterilised.

Pusan, on Korea's southern coast, had a smaller influence field, and 43 were cleared, although a Japanese destroyer, acting as a guinea-pig (or displacement sweep against pressure mines) found a further one during the operation, and sank.

Tsushima Straits, lying between Kyushu and Korea, posed a difficult sweeping problem. Some 6,000 mines had been sown in an area covering 4,500 square miles, and American 'Raven' Class AMs swept there with Japanese sweepers, cutting 3,200 mines in six months of hard work which claimed the last American AM casualty of the war, *Minivet,* which went down at the turn of the year after striking a mine.

These were the main inshore sweeping operations, but there were offshore clearance operations as well. We have already covered one, code-named 'Arcadia'; there were four others, running through to the end of 1945, all in the Yellow Sea. They were code-named 'Klondike', covering 1,300 square miles and clearing 410 mines; 'Reno', covering as much ocean for 620 mines; 'Rickshaw', covering 2,500 square miles in which a big field of 5,000 moored mines had been laid, over 500 of which were cut and destroyed; and finally 'Skagway', covering an even larger area of sea (3,500 square miles) and a fine comparison with the British North Sea situation, with 900 mines cut from an original field of nearly 4,000.

The sheer scale of these clearance operations is most impressive. In 20 separate operations, fields originally containing 32,000 mines were swept of their remaining 11,200. In these operations, American sweepers took the predominant role, although Japanese sweepers were also working in numbers under American control, especially in areas where moored contact fields had been laid.

A further operation was carried out in the last four months of 1945, to open up the river to Shanghai. The M/S Group, designated TG 73.2, included six AMs and 18 YMSs, with a survey unit and six destroyers in support. The group swept in to the river estuary for moored, magnetic and acoustic mines. Damage to sweeps was still a problem here, and two destroyers dashed in with replacement gear from Okinawa; once again, the lack of suitable maintenance and supply ships was keenly felt.

The weather was bad and the currents strong, so the sweeper crews were under constant strain in these dangerous waters. Typhoons hit them, as well as just plain gales, but in two months of hard sweeping they cleared 569 Japanese moored mines and three US magnetics, using 743 sweeper days, of which 532 were from the little YMSs. Inside the river itself only three mines were lifted, in 2,000 ship-hours of sweeping, and LCPRs swept in ahead of the YMSs here, tackling the shallow water where Allied magnetic mines had been laid.

A final but real hazard which must be mentioned in covering Pacific clearance operations, was the typhoons which struck in the middle four months of 1945, and wrought havoc among the sweepers. A total of 12 YMSs were lost by natural causes in this area between September and October, of which ten were lost to typhoons at Okinawa—YMSs 98, 146, 151, 275, 341, 383, 421, 424, 454 and 472. At the northern tip of Okinawa, a typhoon refuge was constructed in an inlet named Untenko; but even here the sweepers were not safe, and just outside that harbour YMSs 341 and 421 were overwhelmed, trying to turn round against the storm into the estuary.

Larger sweepers also felt the storms; the minesweeper flagship *Terror,* just returned from repairing her Kamikaze damage from Okinawa, was damaged yet again, while two of the old 'four-stacker' destroyer minersweepers were lost, after all their faithful war service.

Chapter 9

Sweeping problems in 1945

By early 1945 the cunning of the mine designers (both German and Anglo-American) seemed to be outstriding the effectiveness of the sweeps devised as antidotes—and this was not restricted to the problems of sweeping the pressure mines.

When German mining officers were interviewed after VE-Day, it was found that two fields had actually been laid with mines having arming delay clocks running up to 200 days, while others had 'interrupter clocks' which rendered the mines passive for 24 hours while the sweepers hunted for them, and then active again when the convoys were passing over! It was certainly a good thing for Britain that the Germans did not have the air freedom to lay minefields by aircraft in 1945, as they had done so freely in 1940.

Pulse delay mechanisms (the 'ship clickers') were by then running for up to 15 'clicks', and mines laid in calm water had a very long life. A field of British moored magnetic mines, laid in 300 fathoms in the Denmark Strait in 1942, were still there when the Fortieth and Forty-second Flotillas of 'Catherine' Class fleet sweepers cleared them four years later, in spite of all that Icelandic gales and drift ice could do.

Mixed fields were perhaps the biggest single clearance problem facing the sweepers. The British air mining campaign relied heavily on these—a typical mixed field in 1945 would consist of the following: Double contact magnetic with arming clock—13 per cent; Double contact magnetic with PDMs 27 per cent; Acoustic (normal sensitivity) 13 per cent; Coarse acoustic 13 per cent; Overlap magnetic-acoustic—ten per cent; Magnetic and low-frequency acoustic—seven per cent; Acoustic-magnetic, with PDMs—17 per cent. Mixed in with this little lot for good measure, would be moored contact mines and sweep obstructors as necessary!

To show how complex minesweeping had become, here is the British guide table for sweeping these mixed fields in 1945 and onwards:

Formation	Mine for clearance	Laid by	Swept by	Using sweeps
I	Snag line contact	Brit, Ger	LCT, LCP, ML	Snag line sweep Type D
G	Shallow moored contact	Brit, Ger	ML	Oropesa Mark 7
Single ship	Acoustic—build-up or coarse	Ger	MMS, ML	Explosive sweep, Mk I
Single ship	Plain acoustic	Brit	BYMS, MMS	SA Type C Mk IV

Formation	Mine for clearance	Laid by	Swept by	Using sweeps
G	Moored contact	Brit/Ger	Fleets, BYMS	Oropesa sweep
J				
+ R or S + P or Q	Magnetic-acoustic	Brit	Fleets BYMS, MMS	Mod L or LAA + SA Type C Mk IV
R or S	Sequence mines	Brit	Fleets BYMS, MMS	2 formations using Mod L
P or Q	Magnetic-acoustic	Ger	Fleets BYMS, MMS	LL + SA Type A Mk V + Explosive sweep
R or S	Magnetic-subsonic acoustic	Brit	Fleets MMS	Mod L or LAA + SA Type C Mk I
Single ship	Sonic-subsonic acoustic	Brit/Ger	Fleets MMS	SA Type C + SA Type A Mk IV
P or Q	Acoustic-pressure	Ger	Fleets BYMS, MMS	SA Type A Mk IV + Explosive sweep + Delta sweep with SA Type D Mk II
	Magnetic-pressure	Ger		Delta—Egg Crate

Overlap mines were another British refinement which caused grief among the sweepers. These were mines in which both the magnetic and the acoustic units were alive at the same time, whereas in German mines one unit triggered the other. The danger of these mines was that the standard LL tail and SA hammer box could be rendered useless against them; and while any ship could protect herself against these mines by operating her hammer box continuously, clearing them would be a different matter. The solution seemed to be to use a restricted magnetic and acoustic field, so that the sweeping device should 'look' like a ship to the mine. So it was necessary also to pulse both the magnetic and acoustic fields together, at intervals greater than the passive period of the mine.

At first, a single L sweep with a short live tail of 50 yards was tried, with a low-output hammer box towed much further astern; but later, low-frequency, low-speed pipe noisemakers were hung from the middle of the live section of the tail. The passive period of some of these mines was sometimes as long as five minutes, so it really was a complicated problem.

Three different sweeping schedules were then kept. In the first, ten passes were made over the area, using the sweep described above, but also including the explosive sweep for the first three passes, together with a towed oscillator in the first sweeping sub-division to enter the area. Next, another three passes were made, using LA or LAA sweeps, with hammer boxes being used throughout the formation to catch any remaining convoy mines or plain magnetics. Lastly, a standard LL sweep was used, with a pulsed 27-inch hammer box in each sweeper, to catch the latest German magnetic-acoustic mines, though even then, some anti-*Sperrbrecher* mines might still be missed.

So a different sweeping technique needed to be used for each type of mine, and further nasty surprises were in store in the German mining armoury at the time of VE-Day. One mine, which fortunately did not appear at sea, had a triple unit; a magnetic device actuated the pressure unit, this in turn operated the acoustic unit, and if this, too, reacted to the sweep (or ship), the mine would fire. A very long way from a wire sweep in the horny hands of fishermen in 1939!

Lastly, the differing lifespans of the mines had to be calculated, dating from the time the mines were thought to have been laid. Many ground mines, laid by both sides in Europe and in the Pacific, were fitted with sterilisers to kill the mine after the calculated period—usually a few months. These sterilisers were designed to short-circuit the firing battery, but many failed to operate, and some sank ships long after the war.

The most frequent 'natural cause' in ending the life of a mine would be battery exhaustion. In an acoustic mine, this would take about one year, due to microphone drain. In a magnetic mine, though, it could take as long as four years, after which the battery itself would collapse. Pressure mines usually failed after not longer than two years, when the synthetic rubber in the pressure unit gave way; but if they were pressure-magnetic mines, they could easily go on for a further two years as plain magnetics.

But as often as not, ground mines would sink in mud, or be covered by silt or sand as the tides and currents shifted. If they were acoustics, they would then be regarded as having 'dirty ears'. But they could just as easily be uncovered again later, becoming a fresh menace.

Lessons learned

In 1945, the Royal Navy looked back to 1939 and listed some of the lessons learned in six years of hard minesweeping. Here are some of their conclusions—and no doubt several of them will ring familiar bells in America, too!

1 The principal danger was that of being unprepared—in numbers of sweepers, in trained personnel, and in up-to-date research.

2 The minesweeping forces were largely regarded as a 'private navy'. But ignorance of the sweepers' methods and problems among officers in the larger ships of the fleet was regarded as deplorable. This was balanced by the great enthusiasm of the sweeper crews, who largely disregarded official handbooks of any kind.

3 The plain moored contact mine was still very effective in 1945—largely due to its heavier, unsweepable chain moorings, and the efficient sweep obstructors laid with it.

4 There was a practice, widely used, but to be avoided whenever possible, of using a ship as a minesweeper during the forenoon, and as an anti-submarine escort vessel during the dog watches—resulting in loss of efficiency in both tasks.

5 Fleet sweepers were not employed as influence sweepers until the Normandy landings, and only then was it found that they had serious defects in their generators, causing many operational delays.

6 For moored minefields in clear water, such as the Mediterranean, air spotting could be very effective.

7 Where one's own side hopes to assault, at a later date, an area where a minefield is to be laid, it is a grave mistake not to fit the mines to sterilise before the assault.

8 The use of minesweeping headquarters ships and maintenance and supply ships should be greatly developed. Those in service with the Royal Navy towards the end of the war transformed the coordination, maintenance, and equipment supply situations.

9 Paravanes were the only means of self-defence which large ships had against moored mines; but they were hard to handle in heavy seas, not very effective, and usually unpopular with the crews.

10 Mines in future may well have atomic-fission charges. How will the minesweepers cope with them?

11 Mine location was seen as the probable final answer, especially in shallow waters.

At war's end, it seemed entirely possible to blockade a country, or defeat an assault, simply by the use of unsweepable mines, new types of combined mine, or just mixed fields. The pressure mines in use in 1944 and 1945 showed just how close that situation really was. So, if such a problem were encountered, or if sweeper losses were to rise to unacceptable proportions, mine location seemed to be the answer.

The Royal Navy had tried it out from time to time during the war, so the technique was not entirely new. In the darkest days of the magnetic and acoustic mining campaigns in home waters, the *Vernon* Mine Recovery Flotilla anchored over suspected mines, and divers sought out a choice specimen to recover and dissect.

Then, during the recapture of the North African ports, a contact bar sweep was used in docks and basins. By the time the pressure mines came along in 1944, this process, of divers and special sweeps for location and recovery, was already seen as the potential answer to the unsweepable mine. Once located, it could either be recovered, swept or just blown up.

In the clearance of Le Havre in October 1944, these methods were tried but reported to be a 'conspicuous failure'. One sweep, called the 'locator snag sweep', had 180 tails fitted to a wire A sweep, and dragged them over the seabed to detect metal objects. In another, a spar 31 feet long was slung below a small vessel, and 16 tails from the spar again trailed along the bottom.

Asdic (anti-submarine detection) gear was also tried, with the ship running over any obstacle found and marking it on her echo-sounding recorder.

Nets were also used, again called 'locator snag sweeps', Mark III. The nets were used to 'scrub' suspected areas of seabed, but usually, although they found many strange objects, they found no mines!

A combination of these methods was used in the clearance of ground mines in the swept channel off Ostend in 1945. A Mine Location Flotilla was fitted out, carrying all the known types of gear and sailed to search the bed of the North Sea; they were quite successful, too.

The United States Navy was at least as far advanced as the Royal Navy, probably more so, since the beaches at the Pacific island assaults were thickly infested with mines, and units of frogmen based on landing craft were used consistently and successfully in these operations.

Their work in locating and recovering ground mines was also well advanced, and one of their more effective methods rejoiced in the name of 'Electrical Discontinuity Discriminator'. This was, in fact, a most effective induction-type locator, housed in a towed hydrofoil. Another was the 'Underwater Object Locator', produced by General Electric. This was based on a cathode tube display, and searched a small area of the bottom ahead of the ship, usually an LCT. The supersonic acoustic system was reported to be very complex, but some encouraging results were obtained.

Onwards to the year 2000

So the minesweepers entered the second half of the 20th century.

Mine clearance was completed in the years after the war, but the last messages from 1945 were clear—it had become possible to lay well-designed minefields, with a mixture of anti-ship and anti-sweeper mines, which might well prevent effective clearance. Arming clocks and pulse delay mechanisms, too, were approaching a level of sophistication which might well defeat the sweepers.

Since that time, Europe has enjoyed the longest period of peace in its stormy history; but it is interesting to see that, unlike the period from 1919 to 1939, minesweeping techniques have been quietly but effectively developed.

In minesweeper design, the navies have shown some difference of opinion. In the Royal Navy, no new fleet minesweepers were built after the 'Algerines', but a large class of coastal minesweepers was built in the 15 years after the war. Based on the American YMS design, and of wooden construction but with higher freeboard, and fitted with both wire and influence sweeps, this class formed the basis of the postwar mine counter-measures force. Through the NATO organisation, many units of this class were built for western European navies in the same period, often in the countries where they were to be located.

The United States Navy followed suit, but in addition built a new class of ocean minesweeper (AM), and also provided for some new 'guinea-pig' type ships against pressure mines.

In both navies, mine hunting has increased greatly in importance over the years, as the best method of locating and then destroying ground mines, especially those of the pressure variety. So numbers of the coastal minesweepers were converted to a minehunting role, and were rated as such, while the techniques of minehunting have been constantly developed. The use of newer and specifically-designed Asdic (Sonar) equipment was an integral part of this, as was the use of divers from the minehunter to locate the mines.

In later years, as industrial techniques developed, the Royal Navy has produced a new class of counter-measures vessel, based on the coastal minesweeper design, but with the hull constructed of glass-reinforced plastic. As with any new material, the cost of the first unit was exceptionally high. But as the need to have a non-magnetic hull is still paramount, the use of new materials which could be developed for mass-production in wartime, certainly needed to be explored.

A further project was to endeavour to get agreement on a standard design of mine counter-measures vessel, which would be accepted by, and built by, all the NATO navies. The coastal minesweepers of 20 years ago now need replacing, and a large number of new minehunters needs to be built. There is debate among the NATO partners about the merits of the wood-laminated hull against the newer glass-reinforced plastic; but the economies of a standard design would be significant. In both refitting and in the carrying of stocks of spares, flexibility and economy could be achieved.

But however that debate may go, it is interesting to see that the momentum is being kept up, the need to have the nucleus of an effective mine counter-measures force is clearly seen, and the lessons of unpreparedness before 1939 have been taken to heart.

Appendices

Appendix 1

Profile drawings by Duncan Haws of the main classes of minesweepers (drawn to a scale of 1 in.:75 ft.).

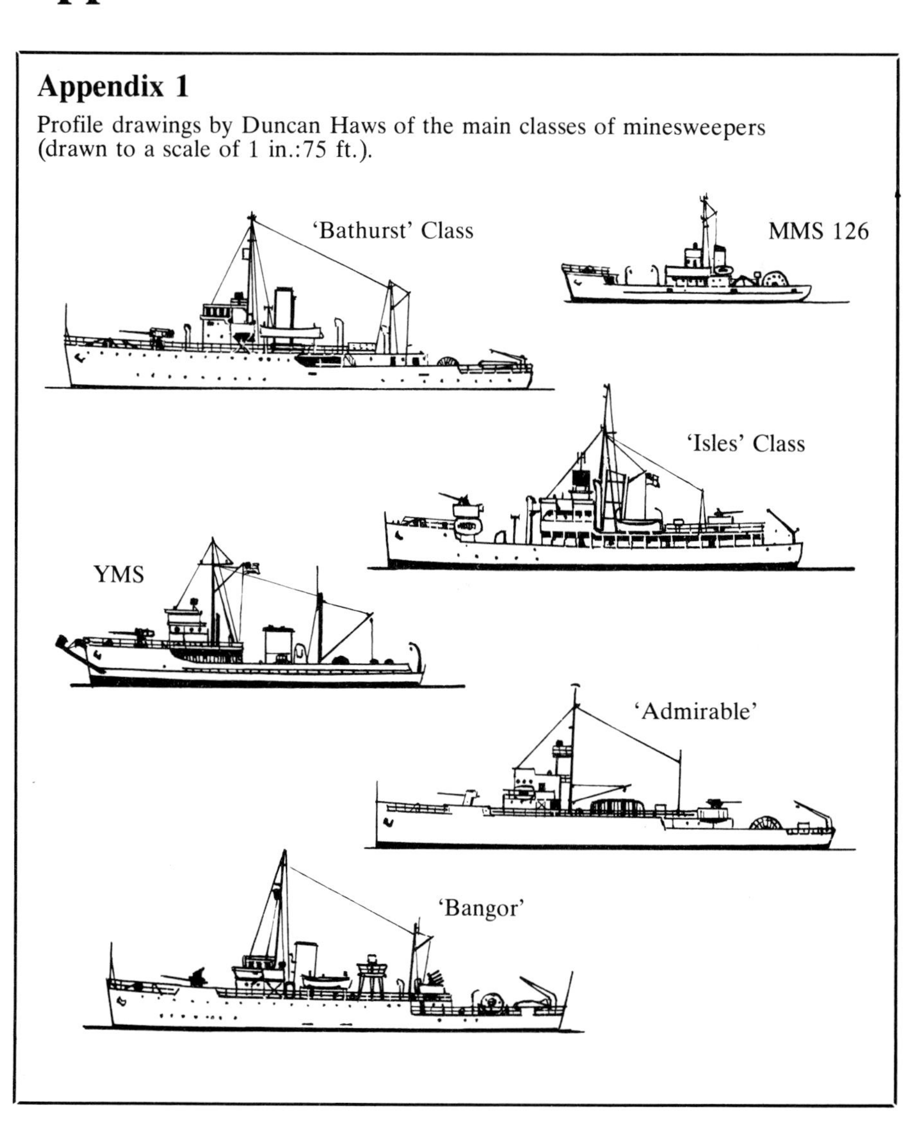

Appendix 1 continued

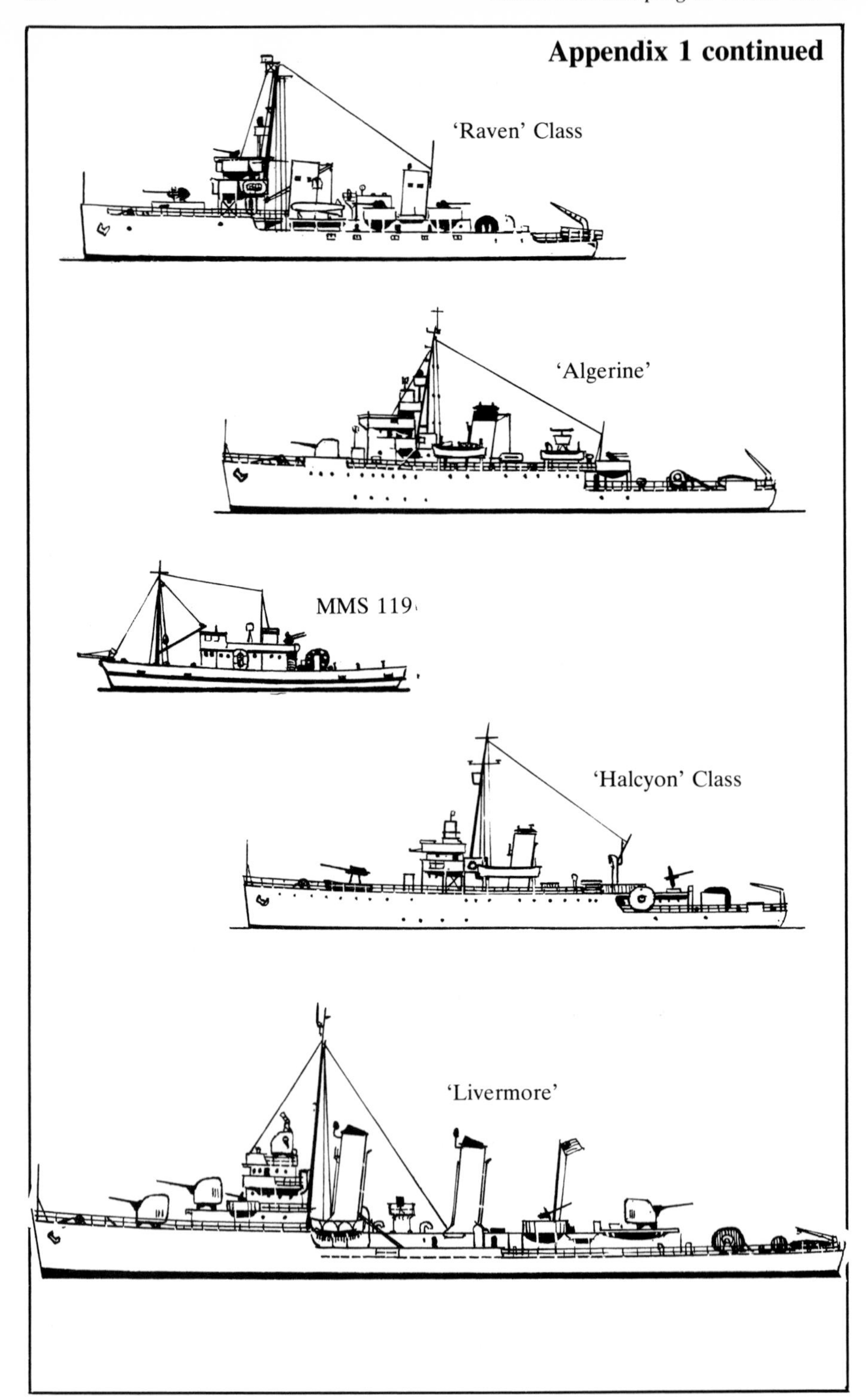

Appendix 2

Increase in minesweepers in commission in six years of war

Royal and Commonwealth Navies

Home waters	**1939**	**1940**	**1941**	**1942**	**1943**	**1944**	**1945**
Class of ship							
'Town'	4	11	10	9	9	7	0
'Halcyon'	17	15	15	13	9	8	7
'Flower'					20	18	
'Bangor'			19	33	25	44	47
'River'				5			
'Algerine'				5	16	29	35
'Catherine'					5	18	19
MMS			45	95	185	220	229
BYMS				2	24	81	92
Paddle sweepers		25	20	6	5		
MD Ships		9	4	2	2		
M/S-A/S trawlers			41	54	51	65	41
'Isles' danlayers						9	29
Oropesa trawlers	40	261	262	240	221	247	82
LL trawlers		150	166	187	173	134	33
LL drifters		51	88	103	55	32	9
Skid towers		41	31	30	38	53	2
Totals	61	563	701	804	835	949	623

Notes
1 All figures calculated at September 1 in each year; **2** Figures include Commonwealth ships in the same class; **3** 1945 figures exclude 15 fleet sweepers then in reserve.

Outside UK waters	**1939**	**1940**	**1941**	**1942**	**1943**	**1944**	**1945**
Class of ship							
'Town'	15	10	6	5	5	4	
'Halcyon'				2	4	2	
'Flower'				13	9		
'Bangor'				9	16	54	51
'Bathurst'			15	21	20	58	56
'Algerine'					8	22	48
MMS				29	61	68	64
BYMS					64	67	90
MD Ships			3	5	2		
M/S–A/S trawlers			11	21	70	36	84
'Isles' danlayers						2	7
Oropesa trawlers		80	177	115	71	144	3
LL trawlers		2	58	100	87	84	15
Skid towers				33	19	6	8
Totals	15	92	270	353	436	547	412

Appendix 3

Comparative specifications of fleet minesweeper (AM) classes

	'Town'	'Halcyon'	'Bangor'	'Bathurst'
Originating country	UK	UK	UK	Australia
Used by Navies	RN	RN	RN RCN RIN	RAN
Displacement (tons)	710	815–875	590–656	650
Length overall (feet)	231	245	162–180	186
Breadth (feet)	28	33½	28½	31
Draught (feet)	7½	7–8	8¼	8½
Main engines	Steam	Recip or turbines	Diesel recip or turbines	Steam
HP	2,200	2,000	2,400	1,750
No of shafts	2	2	2	2
Speed free (knots)	16	17	16	16
Complement	75	80	60	60
Armament				
Main	1 × 4″	1 × 4″	1 × 12 pr	1 × 3″
Secondary	2 × 2 pr	1 × quad. 0.5″ 2 × 20 mm	1 × 2 pr	2 × 20 mm

Appendix 4

Comparative specifications—coastal minesweepers

	'Isles' Class Trawler	MMS 105 ft	MMS 127 ft	YMS	PCS	ML	SC
Originating country	UK	UK	UK	USA	USA	UK	USA
Used by Navies	RN RCN RIN RNZN	RN RCN Allied	RN RCN Allied	USN RN	USN	RN RCN RNZN	USN
Displacement (tons)	545	165	255	215	267	73	95
Length overall (feet)	164	119	140	136	175	112	111
Breadth (feet)	27½	23	26	24½	23	18	17
Draught (feet)	10½	9½	10½	6	7½	4	6

'Flower'	'River'	'4-Stacker'	'Livermore'	'Raven'	'Admirable'
UK RN	UK RN RCN	USA USN	USA USN	USA USN RN	USA USN Russia
950	1,370	1,100	1,620	890	650
205	301	314	348	221	184
33	36	31½	36	32	33
13	14	11½	10½	10¾	9¾
Steam	Steam	Steam	Steam	Diesel electric	Diesel
2,750	5,550	26,000	50,000	3,500	1,710
1	2	2	2	2	2
16	20	32	37	18	15
85	107	140	250	100	100
1×4″ 2 - 6 × 20 mm 2 - 4 × 40 mm	2×4″	3×3″/50 4×20 mm	4×5″/38 2×40 mm 40 mm and 20 mm	1-2×3″/50 Various 3-5×20 mm	1×3″/50 2-4×40 mm 2-4×20 mm

	'Isles' Class Trawler	MMS 105 ft	MMS 127 ft	YMS	PCS	ML	SC
Main engines	**Steam**	**Diesel**	**Diesel**	**Diesel**	**Diesel**	**Petrol**	**Diesel**
H.P	850	500	500	1,000	2,880	1,200	2,400
No of shafts	1	1	1	2	2	2	2
Speed running free (knots)	12	11	10	12	20	20	20
Complement	40	20	21	50	80	16	28
Armament							
Main	1×12 pr			1×3″/50	1×3″/50	1×3 pr	1×3″/50
Secondary	3×20 mm	2×0.5″	2×20 mm	2×20 mm	2×20 mm	1 × 20 mm	3 × 20 mm

Appendix 5

Minesweepers in service 1939–1945

Class	UK	Australia	Canada	India	NZ	South Africa	USA	Totals
A								
Fleet sweepers (AM)								
'Town'*	26							26
'Paddle'*	39							39
'Halcyon'*	21							21
'Flower'	48							48
'River'	30							30
'Bangor'	60		54	4				118
'Bathurst'		60						60
'Algerine'	97							97
'Catherine'	22							22
'Bird'*							22	22
'4-stacker'*							18	18
'Livermore'							24	24
'Raven'							95	95
'Admirable'							124	124
Totals	343	60	54	4	0	0	283	744
B								
Coastal sweepers etc								
Trawlers								
(a) Prewar	910	11	13	1	8	45	60	1,048
(b) War-built	209		12	22	17			260
MD Ships	9							9
MMS								
(a) Short	184		10	5				199
(b) Long	90							90
BYMS	146							146
M/S ML	40							40
ZZ craft	24							24
Stirling craft	2							2
Egg Crates	10						7	17
Concrete barges	2							2
'Bangor' Class tugs	5							5
SGB	6							6
AMc							113	113
PCS							18	18
YMS							481	481
Totals	1,637	11	35	28	25	45	679	2,460

Class	UK	Australia	Canada	India	NZ	South Africa	USA	Totals
C								
Overall totals								
Fleet sweepers (AM)	343	60	54	4	0	0	283	744
Coastal sweepers etc	1,637	11	35	28	25	45	679	2,460
Totals	1,980	71	89	32	25	45	962	3,204
D								
Classes built in more than one country								
'Bangor'	54		60	4				118
'Algerine'	48		49					97
'Raven'							117	117
MMS								
(a) Short	135		64					199
(b) Long	50		56					106
YMS							561	561
Egg Crates	10						7	17
Trawlers	201		20	22				243
Totals	498		249	26			685	1,458

Notes

1 '*' = built before 1939.

2 Some sweeper classes were withdrawn during the war for other duties, eg, RN paddle sweepers, 'River' Class frigates and Mine Destructor ships, and USN PCSs.

3 Numbers shown for the following classes are only those fitted for minesweeping, not the total built of the class: RN 'Flower' Class corvettes, 'River' Class frigates and M/S MLs.

4 Some smaller classes have been omitted from this table, where only a relatively small number were used for minesweeping, and where the numbers so used are not known: RN MFVs, HDMLs and Landing craft, and USN SCs and Landing craft.

5 The totals of trawlers shown are only those fitted and used for minesweeping, and do not include large numbers of others, especially in the RN, used for anti-submarine and other duties. Trawlers were sometimes switched from one duty to another, by refitting; a separate table shows more detail of this.

6 The number of Egg Crates finally built in the UK and in the USA is not known. The numbers shown are an estimate of those constructed in the early stages of the crash production programme in 1944.

7 The 'Bangor' Class tugs are shown separately here, for a more complete pressure-minesweeper picture; but they were originally built as normal fleet sweepers of the class, and are included under that heading also.

8a The tables do not attempt to show where sweepers were switched between countries during the war years, such as UK-built 'Bangors' to India, after some war service, and USA-built 'Admirables' to Russia, also after some service.

Appendix 5 continued

8b But sweepers built in one country, and delivered to, and retained by, another, are shown, as: Australia—'Bathursts' built for RN; Canada —'Algerines' and 'Bangors' built for RN; MMSs built for RN and Russia and trawlers built for RN; United States—'Catherines', Egg Crates and YMSs built for RN.

9 Table D shows the totals of the main minesweeper classes, as noted under **8b** above.

Appendix 6

Comparative war losses of minesweepers

A Fleet minesweepers (AM)

	Theatre of war			
Class	**N Europe and Atlantic**	**Mediterranean**	**Pacific and Indian Ocean**	**Total**
Royal Navy				
'Town'	4	5	0	9
'Halcyon'	9	1	0	10
'Flower'	0	6	1	7
'Bangor'	1	5	0	6
'Algerine'	3	4	2	9
'Catherine'	4	0	0	4
RN Totals	21	21	3	45
Royal Australian Navy				
'Bathurst'	0	0	4	4
RAN Totals	0	0	4	4
Royal Canadian Navy				
'Bangor'	4	0	0	4
RCN Totals	4	0	0	4
United States Navy				
'Bird'	0	0	5	5
'4-stacker'	0	0	7	7
'Raven'	2	4	3	9
'Admirable'	0	0	1	1
'Livermore'	0	0	4	4
USN Totals	2	4	20	26
All fleet sweepers				
Totals lost	27	25	27	79

B Smaller minesweeper classes

Class	N Europe and Atlantic	Mediterranean	Pacific and Indian Ocean	Total
	Theatre of war			
Royal Navy				
Paddle sweepers	10	0	0	10
MD Ships	3	0	0	3
MMS	27	6	1	34
BYMS	5	1	0	6
Trawlers				
(a) War-built	10	3	0	13
(b) Requisitioned	170	40	0	210
Skid towers	16	1	0	17
Stirling craft	2	0	0	2
Egg Crates	1	0	0	1
Concrete barges	2	0	0	2
RN Totals	246	51	1	298
Commonwealth Navies				
RAN trawlers	0	0	2	2
RCN trawlers	1	0	0	1
SANF whalers	0	3	0	3
Allied Navies				
YMS	0	5	0	5
Trawlers	11	0	0	11
Commonwealth and Allied Navies				
Totals	12	8	2	22
United States Navy				
YMS	3	3	26	32
USN Totals	3	3	26	32
All smaller classes				
Totals lost	261	65	29	353

C All minesweeper classes

Totals lost	288	88	56	432

D Overall totals by Navies of minesweepers lost

Navy	Fleet	Coastal	Totals
Royal Navy	45	298	343
United States Navy	26	32	58
Allied Navies (Occupied countries)	0	16	16
Royal Australian Navy	4	2	6
Royal Canadian Navy	4	1	5
South African Naval forces	0	3	3
Royal New Zealand Navy	0	1	1
Overall totals	79	353	432

Appendix 7

Displacement sweeps (against pressure mines)

All Royal Navy, except Egg Crates, designed in United States

Steam Gunboats

Displacement		260 tons
Dimensions	length o/a	145 feet
	breadth	20 feet
	draught	5½ feet
Machinery		Steam geared turbines, SHP 8,000
Shafts		2
Speed (knots — free		30
Complement		27
Armament	main	1—single 3 inch
	secondary	Several double and single 20 mm Oerlikons

'ZZ' craft

Wooden lighters, fitted with LL sweep, and tested against 'Oysters'

Displacement		360 tons
Dimensions	length o/a	145 feet
	breadth	30 feet
	draught	25 feet
Machinery		Gray diesels, SHP 200 = 7½ knots

'Stirling' craft

Two ships built

Displacement		3,980 tons
Dimensions	length o/a	361 feet
	breadth	65 feet
	draught	25 feet
Machinery		Nil

Egg Crates (US Navy designation 'Y' craft)

Displacement and dimensions are not available in records, but former was probably about 2,500 tons, and overall length about 150 feet. They had no machinery.

Concrete barges

Displacement		3,000 tons
Dimensions	length o/a	190 feet
	breadth	40 feet
	draught	18 feet
Machinery		Nil

Tugs for displacement sweeps

'Bangor' Class fleet minesweepers were converted and used for this specific purpose, but the following class was also used.

Appendix 7 continued

'Bustler' Class rescue tugs

Displacement		1,120 tons
Dimensions	length o/a	205 feet
	breadth	38½ feet
	draught	12½ feet
Machinery		Diesel engines, BHP 4,000
Shafts		2
Speed (knots)—free		16

Appendix 8

Some small craft used as assault sweepers or in rivers

		LCT	LCVP	HDML	MFV
Displacement (tons)		200	11	50	114
Dimensions	length o/a	187	36	72	75
	breadth	39	10½	15	20
	draught	4	3	5	9
Machinery		Diesels in all cases			
Speed (knots)		10	9	11½	8½
Complement		12	3	10	10

Appendix 9

Mines laid and swept

The numbers which follow can only be regarded as very approximate; surprisingly, official records do not show these figures clearly in this form—individual commands and flotillas recorded mines swept from month to month, although there were large gaps in this coverage, too. But it is clearly desirable in a history of this kind to include figures showing the overall size of the campaigns.

As will be seen from the narrative, the numbers of mines swept were much lower than those laid. Moored mines broke from their moorings in gales, or sank to the seabed—where even today they can represent a hazard to fishing vessels. Ground mines were easily covered in silt and sand in the changing currents, and so rendered harmless, while their batteries and rubber perished after a few years. Many other mines were set to sterilise themselves after a fixed period.

But of those swept, even half a dozen could require an enormous effort by large numbers of well-fitted and experienced sweepers, before they were cleared.

	Laid			**Swept**		
	Moored	**Ground**	**Total**	**Moored**	**Ground**	**Total**
Northern Europe	97,000	23,000	120,000	2,569	5,924	8,493
Mediterranean	30,000	25,000	55,000	7,460	663	8,123
Pacific	30,000	21,500	51,500	6,000	6,000	12,000

In addition, 752 British moored mines were cleared in northern Europe, and a large number of American mines were cleared by American sweepers in the Pacific. The figures do not include accurate numbers of moored mine obstructors—the Germans were known to have laid at least 30,000 of these in northern Europe alone.

Bibliography

Primary sources

These were found in the Public Records Office in London, and represent a selection only.

General history

Minesweeping Summaries (weekly): (1—10077 and various).
Staff Monograph on British Minesweeping, 1939–45: (189—130).
Technical History of British Minesweeping, 1939–45: (189—129).
Technical History of Mining, 1939–45: (189—120 and —125).

Technical

'Bangor' Class sweepers to fleet tugs: (1—18761).
Combating Oyster Mines: 1—17110).
Egg Crates—Operational Instructions: (1—18739).
Explosive Methods of Minesweeping: (1—18715).
Explosive Sweep Mark I: (1—18720).
Fitting of RDF in M/S trawlers: (1—15358).
LL and Modified L in fleet sweepers, 1945: (1—18771).
Magnetic field of LL sweep, 1940: (212—35/37).

Types of mines

Appreciation of German delayed-action mines, 1940–45: (1—18711).
Description and photographs of German mines, 1946: (213—62).
German pressure-magnetic mine, 1946: (213—167).
German pressure mines, 1945: (189—147).
Japanese mines, 1945: (189—153).

Minesweeping tactics

Clearing anti-tank mines in shallow waters, 1947: (213—634).
Manual of minesweeping, 1929: (186—466).
Mine countermeasures and research, 1946: (213—97/98).
Minesweeping Appendix: (186—557).
Minesweeping in shallow waters, 1943: (1—12932).
Minesweeping Instructions: (212—34).
Operational implications of Oyster mines, 1944: (219—143).

Types of minesweeper
British and German sweeping compared, 1945: (1—18731).
Co-ordination of shipbuilding, UK & Canada, 1943: (1—13030).
Description of mine countermeasures, 1946: (213—97).
Disposal of skids and towing vessels, 1943–45: (1—18728).
Long MMS—trials: (1—15339).
Mine destructor ships, 1942: (219—7).
Paying off paddle sweepers, 1943–44: (1—15360).
Requirements for danlayers: (1—18732).
ZZ craft—trials 1945: (1—18710 and 213—470).

Operations
Mediterranean General minesweeping reports: (199—445—7, 774, 797—8, 806—10, 1428, 1438, 1481—3); *Adriatic* (1—18725); *Aegean* (199—257); *Avalanche* (199—861—2); *Dragoon* (199—120, 863, 865—70, 909—912); *Husky* (199—858—60); *Torch* (199—529, 852, 854, 1—12179).
Pacific General minesweeping reports: (199—1070—76, 1117, 1493—1529, 1540); *Guadalcanal* (199—1324, 1329); *Iwo Jima* (199—1068); *Marshalls* (199—1361); *Okinawa* (199—79, 85—91, 104—7, 460—1, 1061—7, 1530—39); *Philippines* (199—1494, 1498, 1505); *Solomon Islands* (199—1323, 1360).
UK General minesweeping reports: (199—153—6, 182—5, 271—4, 535—51, 741—4, 1167—72, 1350—51); *Nore Command War History* (199—1454—6, 1648); *Normandy* (179—375, 423, 434, 458, 474, 1396, 1558, 1562, 1629, 1646, 199—1445, 1—16745); *Scheldt* (1—18712, 18718, 18737).

British Pacific Fleet
Minesweeping maintenance ships: (1—18724).
Minesweeping requirements: (199—118—9, 152, 1747, 1—18761).

Postwar mine clearance
Clearance organisations: (1—18732, 18150, 18736).
Clearance UK East Coast mine barrier: (1—18772).
50th MSF—Egg Crates: (1—18774).
High-speed LL by SGBs: (1—18792).
Mediterranean clearance: (1—18830—31, 18835, 18855, 1—16357).
N European waters: (1—18742, 18743—46, 18752, 18755, 18784, 18845, 18858—79, 18913).
South-west Pacific: (1—18841).

Secondary sources

Approaches are mined!, The, by Captain Kenneth Langmaid, DSC, RN (Jarrolds, 1965).
Corfu Incident, The, by Eric Leggett (Seeley Service, 1974).
Far Distant Ships, The, by Joseph Schull (The King's Printer, Ottawa, 1950).
H.M. Minesweepers (HMSO, 1943).
Jane's Fighting Ships, 1944–45 (Sampson Low, 1945).
Lilliput Fleet, by A. Cecil Hampshire (William Kimber, 1957).

Most Dangerous Sea, by Lieutenant Commander Arnold S. Lott, USN (ret) (USNI, 1959).
One eye on the clock, by Geoffrey Williams (Macmillan, 1943).
Operation Neptune, by Commander Kenneth Edwards, RN (Collins, 1946).
Royal Australian Navy (2 vols), by G. Hermon Gill (Australian War Memorial, 1968).
South Africa's Navy (W. J. Flesch & Partners, 1973).
Trawlers go to war, by Paul Lund and Harry Ludlam (W. Foulsham, 1971).
United States Naval Operations in World War II, (15 vols), by S. E. Morison (Atlantic, Little Brown).
US Warships of World War II, by Paul H. Silverstone (Ian Allan, 1965).
War at Sea, The (4 vols), by S. W. Roskill (HMSO).
Warships of World War II, by H. T. Lenton and J. J. Colledge (Ian Allan, 1964).

Photographic credits

The author wishes to thank the following for permission to reproduce photographs:

G. A. Osbon
Lt Cdr R. P. Hall VRD RANR
The Imperial War Museum
The Ministry of Defence
The National Maritime Museum
The Public Archives of Canada
The Royal Australian Navy
P. A. Vicary
The United States National Archives
The United States Navy
Wright & Logan

Index

YMS 28 lifting a magnetic mine in her LL sweep inside the harbour at Marseilles, during the clearance of the port in 1944.